DATE DUE

DEMCO, INC. 38-2931

CRACKING THE EINSTEIN CODE

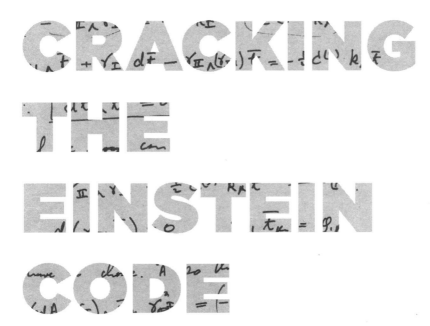

Relativity and the Birth of Black Hole Physics,
with an Afterword by Roy Kerr

FULVIO MELIA

The University of Chicago Press   CHICAGO AND LONDON

FULVIO MELIA is a professor in the departments of physics and astronomy at the University of Arizona. He is the author of *The Galactic Supermassive Black Hole*; *The Black Hole at the Center of Our Galaxy*; *The Edge of Infinity*; and *Electrodynamics*, and he is series editor of the book series Theoretical Astrophysics published by the University of Chicago Press.

The University of Chicago Press, Chicago 60637
The University of Chicago Press, Ltd., London
© 2009 by The University of Chicago
All rights reserved. Published 2009
Printed in the United States of America

18 17 16 15 14 13 12 11 10 09   1 2 3 4 5

ISBN-13: 978-0-226-51951-7 (cloth)
ISBN-10: 0-226-51951-1 (cloth)

Library of Congress Cataloging-in-Publication Data

Melia, Fulvio.
  Cracking the Einstein code: relativity and the birth of black hole physics, with an afterword by Roy Kerr / Fulvio Melia.
     p.   cm.
  Includes bibliographical references and index.
  ISBN-13: 978-0-226-51951-7 (cloth: alk. paper)
  ISBN-10: 0-226-51951-1 (cloth: alk. paper)   1. Einstein field equations.
2. Kerr, R. P. (Roy P.). 3. Kerr black holes—Mathematical models. 4. Black holes (Astronomy)—Mathematical models. I. Title.
✓ QC173.6.M434   2009
  530.11—dc22

                                          2008044006

To

NATALINA PANAIA

and

CESARE MELIA,

in loving memory

CONTENTS

PREFACE

Something quite remarkable arrived in my mail during the summer of 2004. It was an invitation to attend a *fest* in Christchurch, New Zealand, in honor of Professor Roy Kerr's seventieth birthday. The letter triggered an odd feeling, reminiscent of the time I first heard about Galileo Galilei turning down a professorship at Harvard.

The Catholic Church had forbidden Galileo from continuing his scientific work, and seizing upon the opportunity of attracting a major figure of learning, Harvard University set about recruiting him in 1638. Fearing the arduous journey at his advanced age, however, Galileo declined the offer, thus becoming the most iconic figure ever to do so at this esteemed university.

The names Galileo and Harvard were never meant to go together, at least not in my view of history. Galileo was mythical, larger than life, and a giant of antiquity. In contrast, Harvard was young, founded by a band of scrappy settlers in the New World. Did they really overlap?

To the high-energy astrophysicists of my generation, the name Kerr is similarly iconic, perhaps even reaching mythical status, like that of Galileo. Kerr is not so much a person, we thought, but rather the designation of the most famous solution to Einstein's equations of general relativity. Many of the objects we encounter in the cosmos are so condensed that time and space are inextricably entangled by their strong gravity. And the Kerr metric, as his solution is known, is the description of spacetime that everyone uses.

Yet the man Roy Kerr has been a complete mystery to the generation of astronomers and physicists that followed the golden age of relativity, from 1960 to the mid-1970s. So intense was the effort of cracking Einstein's code during that period, that many of the bright, young Turks facing that challenge later chose different intellectual pursuits.

Roy Kerr's presence in the northern hemisphere diminished after his return to Christchurch, but his legend grew as the passage of time magnified the importance of his work. Like the story of the New World reaching out across history and the Atlantic Ocean to touch Galileo and the antiquity he represented, the letter from New Zealand that summer connected me to one of the greatest mathematical physicists of the twentieth century.

It brought into sharp focus the sad truth that although we know much about Einstein, considered one of the greatest physicists of all time, very little has been written about the story of how his theory of relativity has been made relevant to the physical world. Einstein himself was only partially successful in solving his own equations, of "breaking his own code"—a measure of the difficulty of this task.

That is why writing this book has been a labor of love. Getting to know the man who long ago set the stage for most of the research we do has turned into a pleasure I could not have anticipated as a student. Astrophysicists of my generation now have an opportunity of reaching out across the decades and the Pacific Ocean to connect with Roy Kerr the man, and to share with him the dramatic progress being made in the study of black holes. It is rewarding enough to see the glee with which he learns of the impact his work has had.

I am grateful to several individuals whose help was critical to the writing of this book. Roy and Margaret Kerr were exceptionally warm hosts during my many meetings with them, and the kind of accessibility they provided is what turns a simple story into an eyewitness account of a historically important struggle. I am grateful to David Wiltshire and the other members of the physics department at the University of Canterbury, and to the Erskine Foundation, for coordinating my visit to Christchurch, one of the loveliest cities on the planet. My editor, Jennifer Howard, has been an inspirational shepherd for this project, offering her much-needed wisdom at many critical times. And to Marilyn Head, a New Zealand journalist who has written about Roy Kerr for the media down under, I owe a debt of gratitude for her support and for the valuable resources she provided. I also want to thank Andrzej Trautman for releasing the many beautiful images shown in chapter 5. The majority of these have never been published before, but they provide a wonderful glimpse into the people and events that sparked the golden age of relativity.

I am very happy to acknowledge the National Science Foundation, the National Aeronautics and Space Administration, and the Alfred P. Sloan Foundation, for generously supporting my research in astrophysics, specifically on black holes and other compact objects. It is through the work they sponsored that the subject of this book has evolved into one of my abiding interests.

Finally, to Patricia, Marcus, Eliana, and Adrian, and to my parents, whose guidance has been priceless, I extend my enduring love and gratitude.

FULVIO MELIA
Tucson, Arizona
September 2007

1 :: EINSTEIN'S CODE

The scene could have been straight out of Universal Studios, Hollywood. Two men are breathing rhythmically in a smoke-filled modest little room facing south toward the capital of Texas. They sit quietly no more than an arm's length apart, lost in thought. Roy Kerr, the younger of the two, is hunched over a secondhand desk with his back to the door, studying the equations he has just scribbled in a notebook. His older friend and mentor, Alfred Schild (1921–1977), puffs away at a pipe while occupying a worn-out armchair to his right. It is late morning, and rays of sunlight filter through the bushes outside the window, creating a mosaic of light and shadow across the paneled walls.

At stake in this drama is the breakthrough solution to Einstein's equations of general relativity that have defied the greatest scientific minds of the twentieth century. So impenetrable is this description of nature, that Einstein himself succeeded only partially in divining its impact on the meaning of space and time. But it is now 1963, and the freshly minted mathematician out of Cambridge University, settling in at Schild's newly established Center for Relativity at the University of Texas at Austin, is about to crack the great physicist's famous *code*.[1]

Much has been written about Albert Einstein (1879–1955) and his

---

1. As the reader will have realized by now, the use of a *code* analogy for finding an exact solution to Einstein's equations is helpful in appreciating the difficulty of this task, but one should not take this comparison too far. It is important to recall that many individuals attempted to understand the physical meaning of these equations by exploring their behavior in limiting situations, which do not require an exact solution. What is true, however, is that without an exact solution, such as the Schwarzschild metric (see chapter 3) or the Kerr metric that we will learn about shortly, it had not been possible to study the nature of spacetime close to the gravitating objects themselves. Thus, finding these exact solutions was key to unlocking the full power of Einstein's equations, much as cracking a code enables its use for all subsequent applications.

profound influence on our view of the universe, but very little is known about the golden age of relativity, spanning the period 1960–75 following his death. This book is the story of the brilliant young scientists of that era who accepted the challenge of unraveling the mysteries hidden within the seemingly unfathomable language of general relativity, culminating with Kerr's uncloaking of one of the most important and famous equations in all of science.

It is not always possible to discern the reasons why a scientific investigation meanders raggedly or slowly toward its ultimate goal, but in the development of relativity, the complexity of its mathematical formalism is certainly one of them. The difficulty of designing suitable experiments to test Einstein's theory is another. But neither of these reasons emerged for want of interest. Einstein became an instant celebrity soon after founding general relativity in 1915–16, with the quick, auspicious confirmation of one of his predictions—that gravity should bend the path of light as well as that of any particle with mass. This result resounded across the front pages of newspapers around the world, and scientists took note of the new ideas almost right away.

Indeed, only a few months after Einstein's publications began circulating around Europe, Karl Schwarzschild (1873–1916), a soldier on the Russian front, amazingly already succeeded in finding a description of space and time consistent with Einstein's theory, but only for a highly idealized situation, that is, for the gravitational field surrounding a static, spherically symmetric mass. Einstein greeted Schwarzschild's news with enthusiasm, and his solution is used to this day to describe phenomena in regions of strong gravity.

How odd, then, that arguably the most elegant scientific theory ever devised should slowly wither into the decades that followed this remarkable beginning. Those who knew him best have written that already by the 1930s Einstein's interest in general relativity had almost completely lapsed. Having by then moved to Princeton, he could count the number of colleagues working in this field on just one hand. Relativity theory had become irrelevant to science—a situation that sadly persisted up until Einstein's death. He would never know about the breathtaking discovery that would be announced just a few years later—a splendid confirmation of another prediction made several decades earlier.

This experimental achievement—a compelling demonstration in 1960 by the Harvard physicists Robert Pound and Glen Rebka[2] that time slows down in the presence of gravity—sparked the revolution that followed dur-

2. See Pound and Rebka (1960). An image of Robert Pound, Glen Rebka, and their experimental setup appears in figure 4.4.

ing relativity's golden age, leading to that special moment in Roy Kerr's sepia-tinted office shortly afterward.

In the intervening years, failure to uncover practical applications of Einstein's theory was compounded by the lack of progress in the experimental verification of general relativity as the correct description of nature. Ironically, part of the problem was the Schwarzschild solution itself, which in time would be used to predict that truly bizarre objects, variously called *dark* or *frozen stars*, must exist somewhere in the cosmos. Today we call them *black holes*, but back then no one—particularly Einstein—believed they could be real. Yet the Schwarzschild solution clearly demonstrated that the end result of a gravitational collapse must be the formation of a singularity—a point of infinite density—that creates a closed pocket of space and time forever disconnected from the outside world.

Many thought that nature could not possibly create something so unreasonable, believing that no object in the universe is truly static and that, at the very least, its rotation would inhibit any collapse toward a singularity. And so began the search for the "holy grail" of relativity—a description of space and time surrounding a spinning object. Everything we see in the universe rotates, the argument went, so in order to demonstrate that Einstein's theory is a true description of gravity, we must be able to show that his equations do in fact describe space and time surrounding a spinning mass.

But what a challenge this turned out to be! Some of the world's most renowned physicists spent their entire careers working on this problem, making some progress, but losing interest or hope in their waning years. Of course, by the middle of the twentieth century, quantum mechanics had forged well ahead of relativity in relevance and measurability, cementing its place as the overarching theory in the physics pantheon. It didn't help that relativity and quantum mechanics seemed to be incompatible with each other, since the former uses perfectly measurable locations and times, whereas the latter is essentially a theory of spatial imprecision.

The Pound-Rebka experiment changed all that, principally because even the quantum mechanicists could not easily discount its remarkable implications. In fact, among the staunchest supporters of relativity and its relevance to modern physics was Vitaly Ginzburg,[3] co-recipient of the 2003 Nobel Prize in physics for his work in the 1950s on superconductivity, a phenomenon in which some materials carry currents freely, without any resistance, by virtue of a quantum effect that becomes important at very low temperatures.

Though his interests were mainly in quantum mechanics, Ginzburg

---

3. See figure 5.9.

would nonetheless become an inspirational figure to many young physicists drawn to Einstein's theory in the early 1960s. Listening to him,[4] the twenty-eight-year-old Roy Kerr understood that "cracking Einstein's code" was indeed the challenge he should ply with his mathematical talents—a task that would soon bring him to that fateful day sitting next to Alfred Schild in his smoke-filled office.

But the story begins well before Kerr's arrival in Texas, even before Einstein himself, in fact. Musings concerning the nature of space and the meaning of time began to appear thousands of years earlier, in places such as the Greek colony of Elea in southern Italy. Before we explore the evolution in Einstein's thinking that led to his theory of general relativity, and the inspired work that followed during its golden age, we will therefore begin by tracing some of the incipient thinking that led to the problem in the first place. Our journey commences in the fifth century BC with the Greek philosopher Zeno, a man clearly far ahead of his time. Zeno realized even back then that the notion of an absolute space independent of time was paradoxical—anticipating by several thousand years the eventual unification of the two into the structure we now refer to as simply *spacetime*.

4. Figure 5.10 shows some of the participants at the very influential 1962 Warsaw meeting on Gravitation and General Relativity.

2 :: SPACE AND TIME

Physical reality emerges from the sequential defoliation of spatial phenomena. At least that is how we perceive the external world. Never mind that the nature of space and the meaning of time—scrutinized by thinkers from the dawn of civilization in India, China, and Greece, through to the modern era—have defied complete demystification.

Across the ages, many have plumbed the depths of the unknown to uncover nature's secrets. Without question, however, the greatest contribution to our view of the universe was Albert Einstein's general theory of relativity (fig. 2.1), which takes space and time and folds them, twists them, and pulls them into a single interwoven unit, at once alluring and uncomfortably disquieting.

But his code of nature proved to be one of the most difficult to crack in all of science, and it was left to his colleagues and successors to continue the search, occupying them for the better part of the twentieth century. Their story is emotive and at times exhilarating, leaving us with little doubt concerning the remarkable impact this work has had on our cosmic perspective.

This is not say, however, that we finally understand completely what space and time are. Nature works in certain ways, and we can describe its methods, though at the most fundamental level, why things behave the way they do is still unknown. Every civilization, from antiquity to the present, has formulated its own notions of space (the "heavens") and time (the "beginning" and the "end"). Some luminaries, such as the Buddha, Siddhartha Gautama,[1] the spiritual master from ancient India who founded Buddhism in the fifth century BC, and Lao-tzu,[2] the Chinese philosopher credited with writing the central Taoist work (the *Tao Te Ching*) around the same time, commented (or wrote) extensively on this subject. In the Western world,

---

1. See, e.g., Armstrong (2004). **2.** See, e.g., Kohn (2005).

FIGURE 2.1. Albert Einstein in the patent office in 1905, at the time he published his first paper on the special theory of relativity. (The image was provided by the Institute Archives at Caltech, courtesy of the Hebrew University in Jerusalem.)

the Greek philosophers Aristotle, Socrates, Plato, and Zeno[3] had the greatest influence, extending their intellectual reach into the Renaissance, when Galileo Galilei (1564–1642) and Sir Isaac Newton (1642–1727) would supplant their erstwhile inscrutable view of the world with scientifically more consistent theories.

The essence of their teachings, crystallized in our consciousness, forms mental images we still use today. We view space as a three-dimensional continuum that envelops us, while time seemingly flows alone, serenely, unaffected by influences in the physical universe. In combination, they

---

3. See, e.g., Guthrie (1981).

provide the canvas upon which the colors of interaction are splashed, the basis upon which matter, radiation, people, and planets are tangibly manifested.

Firmly rooted to Earth—the origin of all known life—the ancients understandably viewed space as an absolutely position-dependent concept. The Chinese, for example, considered the structure and rhythms of the universe to have perfect unity and continuity, though with an evident center: China in the middle of the world, its capital at the hub of the kingdom, and the royal palace at the nucleus of the capital.[4]

People in the West had similar notions. According to Greek mythology, Zeus charged two eagles with finding the center of Earth, and released one to the east, the other to the west. They met at Delphi, on the slopes of Mount Parnassus in Greece. Over time this would become the site of the most important oracle of the classical world. Here stood the *omphalos* stone, which, translated into English, simply means (Earth's) "navel," revered by the Greeks as the center of Earth and the whole universe.

For a thousand years, up to the time of Galileo and Newton, thinkers who worried about such things also had to contend with Aristotle's concept of an absolute space fixed to the firmament and an absolute time. In his "cosmogonic" system, he argued that change (or movement) were associated with Earth and the Moon because of their imperfections. Perfection was seen as a state with no need for change, a characterization that seemed to apply to other planets, the Sun, and the stars, which were thought to be immutable and eternal. Such an exalted state of being requires an independent, self-defined space and an equally unrestrained and perfect time. This was the worldview Newton inherited as a student at Cambridge, one that over the course of his life he would transform with deep, insightful analysis.

In retrospect, we might wonder why it took so long to address the apparent inconsistencies in the interpretation of space and time head-on. To be honest, however, even today we do not have a compelling answer to some of the difficulties uncovered by the early thinkers, at least not in terms of pre-relativistic ideas.

Take Zeno's argument known as "the Arrow Paradox," for example. Zeno was born around 495 BC in the Greek colony of Elea in southern Italy, but very little is known about him. Though he developed some forty paradoxes, only eight have survived, thanks to the efforts of subsequent writers, such as Aristotle. Zeno's goal was to demonstrate that motion, change, time, and plurality are mere illusions—that in reality the universe is singular and immutable.

4. See, e.g., Eliade (1984).

Like the other arguments known to us, the Arrow Paradox seems illogical, perhaps even confusing, but it cannot be discarded quite so easily. Aristotle himself abandoned Zeno's paradoxes without really proving their (assumed) fallacy, but they were revived many centuries later by mathematicians such as Bertrand Russell (1872–1970) and Lewis Carroll (1832–1898).

Let us suppose, as Zeno did, that absolute space and absolute time exist without any direct connection between the two, and let us imagine an arrow launched (for the sake of argument) along a straight-line trajectory. Then, simply stated, if the arrow exists distinctly at a sequence of discrete instants in time, and if no motion is discernible in any given instant, there cannot be any motion from one instant to the next.

Bertrand Russell described this as "a plain statement of an elementary fact."[5] If we were to make an image of the arrow in flight, it would be very difficult for us to tell from the photograph whether the arrow was moving or not. With only a modest shutter speed, the arrow might appear blurred, since it would have changed location slightly during the exposure. But as the shutter speed increases, and the corresponding interval shrinks to zero, this blurring effect disappears, and in the limit when the interval goes to zero, there would be no difference at all in the appearance of a moving arrow and its stationary counterpart. But if there is no discernible difference between a moving and a nonmoving arrow in any given discrete instant, how does the arrow "know" from one instant to the next whether it is moving? How can the causality be transmitted forward in time through a sequence of such instants, in each of which motion does not exist?

Attempts to rationalize this inconsistency have met with only limited success, and the essential physical question remains—that is, what is different between the moving and nonmoving arrows during any given instant of time? No one now believes Zeno's denial of any real physical motion, but an answer to this question, partial as it is, could only emerge from the physical principles developed during the twentieth century. Modern physics has concluded, along with Zeno, that the classical image of space and time was simply wrong to begin with, and in fact motion could not be possible in a universe constructed according to the old ideas, as we shall soon see.

The evolution away from the ancient concept of space and time actually began with Galileo and Newton, two icons of the classical physics pantheon. In 1687 Newton published one of the most influential books ever written— the *Philosophiae Naturalis Principia Mathematica*, or *Principia* for short. In it he provided a new paradigm, though still describing time as a one-dimensional continuum existing on its own, and space as an absolute three-dimensional

5. See Russell (1996).

substrate, which might or might not contain material things. But Newton introduced a novel idea, subtle at first, but profound in its influence: space, he declared, does not have a center. Nor is it always necessary to describe motion relative to a special, fixed frame. The *Principia* shattered Aristotelian orthodoxy by abolishing the clear demarcation between heaven and Earth. No longer were the stars and the planets immutable embodiments of perfection, establishing the ultimate standard by which all (imperfect) motion ought to be measured.

What led Newton to this intellectual unshackling was the realization that though space could be absolute, an individual could still be moving through it without being able to detect his speed. For the purpose of this discussion, imagine that you are making measurements inside the passenger car of a moving train, and that a second person is making her corresponding measurements in her own car attached to another train moving at a different velocity. According to Newton, all such vehicles are equivalent and equally viable for divining nature's inner workings—as long as they are not accelerating.

Now suppose that all of your windows are shut. Can you nonetheless tell if the train is being accelerated? The answer, of course, is yes! A chain dangling from your hand moves to an oblique position backward when the train is speeding up, and forward when the train is slowing down. The chain hangs straight down when the train is moving at constant speed—no matter what that speed is.

So how does the chain "know" when to hang straight down and when to hang obliquely? Newton posited that there must exist an absolute space, against which acceleration is manifested, but that the laws of nature are not affected by how fast an observer is moving through it. As passengers in your respective trains, you cannot tell how fast you are moving relative to this absolute space, so your (constant) motion must be completely irrelevant. You can only measure your acceleration by examining how far the chain leans away from the vertical direction. In the *Principia*, Newton described the situation as follows:

> Absolute space, in its own nature and with regard to anything external, always remains similar and unmovable. Relative space is some movable dimension or measure of absolute space, which our senses determine by its position with respect to other bodies, and is commonly taken for absolute space.

The *Principia* reigned as the new orthodoxy for over 150 years, while everyone continued to ignore the Arrow Paradox introduced by Zeno more than a thousand years earlier. But natural philosophy, the burgeoning field of physics, was heading toward a crisis. By the middle of the nineteenth century, the Scottish physicist James Clerk Maxwell (1831–1879) achieved a surprising synthesis of all the known facts pertaining to electric and magnetic phenomena. The most astonishing result of all was his demonstration that light itself is an electromagnetic occurrence, propagating at a fixed speed completely independent of anyone's perspective.[6] How could nature do this after Newton had convincingly demonstrated the importance of relative motion between different observers? Though no constant motion could be discerned relative to absolute space, Newton had argued, an individual could certainly measure someone else's speed relative to himself, so they surely could not infer the *same* speed for light. This incomprehensible deviation from Newton's worldview created a state of confusion for the next fifty years.

To see how counterintuitive these results are, consider again your train, now traveling at 180 miles per hour relative to the platform, while your friend's train is coasting along at 70 miles per hour on a parallel track. The second passenger sees you moving past her at a rate of 110 miles per hour. Standing at the station, we decide to flash a pulse of light in your direction once the trains have passed. According to our measurements, not only does the light pulse overtake the two trains at a speed of approximately 670 million miles per hour, but, for reasons that are still unknown, it does so at exactly the same speed relative to you and to your friend, even though she is moving relative to you, and both of you are moving relative to us.

It is quite conceivable that if someone had taken the Arrow Paradox seriously, they might have anticipated the theory of special relativity hundreds of years before Albert Einstein finally grasped the true meaning of light's aberrant behavior early in the twentieth century.

The fact that light travels at the same speed, regardless of who makes the measurements, completely dismantles the classical notions of distance and time. If two people moving past each other claim that light has the same speed, then the distances they measure in ascertaining how far it travels in a given time cannot be the same. Distance is a measure of how much space exists between two points, and a duration corresponds to how much time has elapsed between two events. As the ratio of distance over time, speed is therefore intimately connected with the properties of space and time. Any empirical evidence that defies our intuitive concept of speed, such as the

6. In 1887, following Maxwell's theoretical demonstration that light ought to have a fixed speed, Albert Michelson (1852–1931) and Edward Morley (1838–1923) designed and executed one of the most famous experiments in physics. With it they measured how fast light travels along Earth's orbital motion and transverse to it, proving that the speed is identical in every case.

constancy of the speed of light, must surely be hinting at the fact that the measurement of distance or time, or both, depends on who is making the measurement—odd, indeed.

Newton, it seems, did not go far enough in supplanting the Aristotelian orthodoxy. He was correct in declaring that space has no center; but the constancy of the speed of light must surely mean that absolute space and absolute time do not exist either. Zeno was right after all, though not exactly for the reason he imagined.

Nature is telling us that the Newtonian worldview breaks down when very high speeds are involved, so let us examine how measurements of distance and time must be handled in order to correctly interpret the behavior of light.

In 1905 Einstein wrote a seminal paper that would bring him widespread fame by the end of the following decade (see chapter 4). In it he proposed what is now called *the special theory of relativity*, an attempt to reconcile the surprising discovery that light always travels at a constant speed, with the classical view that the laws of physics should look the same to any observer, moving or not.

Special relativity begins with two postulates: (1) that the laws of physics are the same for all non-accelerated individuals, and (2) that the speed of light in a vacuum is independent of the motion of all observers and sources, and is observed to have the same value. The latter forces us to accept the fact that distances and time must be different for different observers. The former implies something equally profound about space—that we cannot infer an absolute velocity relative to it. Using your hanging chain in the train, you can tell us whether you are accelerating or not. But with the windows closed, you have no way of measuring your speed relative to the ground.

Actually, the second postulate is a much more dramatic departure from classical thought than we have acknowledged thus far, for it requires not only a lack of universality in the determination of physical quantities, but, more importantly, it also demands an *intertwining* of the measured time and space. Time is no longer something we may consider as a river, flowing at a uniform rate, independently of what is happening in space.

Reading Einstein's own words,[7] we may see how the transition away from absolute space took hold:

> I stand at the window of a railway carriage which is traveling uniformly, and drop a stone on the embankment. . . . I see the stone descend in a straight line. A pedestrian who observes the misdeed from the footpath

7. Einstein's book *Relativity: The Special and General Theory* first appeared in 1920, soon after the foundation of general relativity. The most recent edition was published in 2006.

notices that the stone falls to Earth in a parabolic curve. I now ask: Do the "positions" traversed by the stone lie "in reality" on a straight line or on a parabola? . . . We [must] entirely shun the vague word "space," of which, we must honestly acknowledge, we cannot form the slightest conception, and we replace it by "motion relative to a practically rigid body of reference." . . . With the aid of this example it is clearly seen that there is no such thing as an independently existing trajectory, but only a trajectory relative to a particular body of reference.

An important consequence of this result is that simultaneity cannot be universal. Since time flows differently for different individuals, one observer may decide that two events seen some distance away occur at exactly the same time, while a second individual moving relative to the first will see one event happening before the other.

Abandoning the train, you now decide to fly to your next destination, choosing a seat right at the midpoint between the nose and tail of the airplane. You decide to turn on a beacon of light at your position and watch the light pulse move out in both directions, to the front and rear of the fuselage. Since the endpoints are equidistant from you, the light reaches them at exactly the same time, and if you had put mirrors there, the reflected light would reach you at exactly the same time from both directions. This is what we mean by simultaneous events.

But your friend standing on the ground sees something entirely different. As the plane passes overhead, she sees the forward light pulse taking longer to reach the nose than its counterpart takes to reach the rear. Since the plane is moving forward, by the time the light pulse reaches the mirror, the nose has moved beyond the position it occupied when the beacon was turned on. The opposite happens with the tail mirror; the rear of the plane has also moved forward while the light pulse was traveling backward.

Your friend on the ground therefore sees the light pulse reaching the rear mirror *before* its counterpart reaches the front. According to her, the two reflections cannot be simultaneous, but both of you are correct. Differences such as this are unavoidable if the speed of light is always the same, since different observers see events occurring in *different* places and at *different* times.

We must think of time as an objective expression of the changes that occur in the state of matter. Time exists in a certain region of space only because the constituents at that location are changing. We infer that time has passed because the inner workings of the clock have caused its hand to move across its face. Time and movement are inseparable concepts. If all physical processes within a given region of space were to slow down to half their normal speed, time would correspondingly "flow" at half the rate with which it flows elsewhere. As a measure of how quickly things change, time would therefore stop if everything were to become frozen in place.

How intriguing is it, therefore, that in Buddhism the ultimate aspiration is to reach Nirvana, a state in which time ceases to exist? So perfect is this state, that no more change and, therefore, no more time is necessary. Ironically, Aristotle himself had similar ideas about time and change, even though his worldview was clearly absolutist, particularly with respect to space. About time, he wrote that "movement . . . is also continuous in the sense in which time is, for time is either the same thing as motion or an attribute of it."

Special relativity compels us to think of space and time in a unified manner, losing any semblance of division by advancing the terminology to the next level, in which we speak not of space and time but, rather, of *spacetime*. What affects space bears on the change within it and hence also affects time.

This is the reason why lengths contract and times dilate when we compare the measurements of one observer to those of another. Still sitting in the airplane, you wish to measure the distance from your seat to the nose of the fuselage. You prepare the beacon and look at your watch. You turn on the light and wait for it to strike the mirror and reflect back to you, all the while keeping track of the time. You know that light travels at a fixed speed, so you calculate the distance by simply multiplying the speed of light by half the elapsed time (since this is a round-trip).

"Hold on," says your friend on the ground. "The light you release with the beacon has to catch up to the mirror since the plane is moving forward, so the total length traversed by the beam is larger than the half-length of the fuselage." But the speed of light is the same for both of you, and so your friend's watch shows that a longer time has passed for her than for you. The net effect is that a length she measures on the ground appears to be shorter to you, and the time you measure on your watch is longer for her. These two effects—length contraction and time dilation—capture the essence of special relativity.

Wandering about the sun-splashed hills of Elea two and a half thousand years ago, Zeno correctly anticipated the fundamental problem with absolute space and an independent time. In the classical mind-set, there really was no way to distinguish a moving arrow from a stationary one during the briefest possible instant. But the classical concepts are wrong for several reasons.

To begin with, it is physically impossible to determine both the arrow's position and its speed with infinite precision. These quantities are not known a priori; indeed, we are not even aware of the arrow's existence until we sense the sunlight reflected off its surface. However, any interaction of this kind alters the arrow's position and speed, even if only by a tiny amount imperceptible to the human eye. With the benefit of hindsight, we now recognize this as an important element of quantum mechanics, and with it we realize that Zeno's argument was actually ill-posed.

However, Einstein's theory of special relativity answers Zeno's concern even without the benefit of quantum mechanics. Space and time are not independent, it turns out, and the two are coupled by that peculiarly constant speed of light. Any motion alters the conditions for simultaneity, and therefore events observed by someone moving with the arrow do not coincide with those seen from the ground. This distinction alone is sufficient to render the moving arrow distinguishable from its stationary counterpart.

Einstein's resolution of the Arrow Paradox apparently never occurred to Zeno, though we cannot be completely certain. Still, it can be said that special relativity fully vindicates his skepticism and profound intuition about the nature of motion. Of course, the fact that light moves at a constant speed was not known until the nineteenth century. Nonetheless, something like special relativity might have emerged much earlier, simply on logical grounds.

3 :: GRAVITY

In 1907, two years after the advent of special relativity, Einstein began to wonder how his new concepts of space and time could be made to work with gravity. He would later describe the idea taking shape in his head as the "happiest thought of my life." A home owner, falling from the roof of his house, he realized, experiences no gravitational field. Yet he is clearly still in Earth's embrace.

Actually, the stage was set for this revolutionary idea centuries earlier when Galileo's experiments led to the surprising discovery that bodies fall at a rate independent of their mass. By sliding weights down an inclined plane, he was able to eventually convince everyone that a body's mass had nothing to do with the rate at which it fell. According to his biographer Stillman Drake,[1] the Leaning Tower of Pisa story is probably true, only because Galileo would have used such a dramatic, theatrical demonstration to capture the public's attention; he already knew the result from his experiments at home. In this famous anecdote, Galileo dropped several lead balls of different masses from the tower, showing that they all hit the ground at the same instant, even though they were clearly being accelerated by Earth's pull, since they moved faster and faster as they fell.

Today we see direct evidence of this result and its important consequences. We need only conjure up a mental image of astronauts floating freely about their orbiting spaceship. Everything inside the craft is actually falling toward Earth, but because the rate of acceleration is the same for every object, the cabin appears to be "floating." The astronauts sleep in all sorts of positions, feeling no stress on their bodies since their world is *free* of gravity. Yet the space station cannot break away from Earth's influence. How do we understand this?

Well, both Einstein's hapless home owner and the astronauts and all

1. See Drake (1981).

their belongings are indeed all falling toward Earth. However the spaceship is moving sideways with just the right speed (about 17,400 miles per hour) to offset its motion downward. As it falls, the space station keeps up with Earth's curvature, and it therefore follows a circular path, never reaching Earth directly. Above 17,400 miles per hour, the craft would move sideways too fast, breaking free of Earth's orbit. Below this speed, the spaceship would fall to a lower level, even hitting Earth if it had no sideways motion at all.

But if everything experiences the same acceleration due to Earth's pull, then the effect of gravity must be entirely equivalent to a uniform acceleration throughout a given volume of space. Einstein explained it as follows: imagine that you are standing inside an elevator out in space, very far from any object that can generate a gravitational field. Now you and your belongings are in a microgravity environment, but suppose that a rope tied to the top of the elevator begins to accelerate you. Your feet would feel a force "upward" from the floor. Inside the elevator, however, you could not tell whether your environment is accelerating upward, or if you and your belongings were being pulled downward (by gravity). The two effects are entirely equivalent because, in both cases, everything inside the elevator accelerates downward at the same rate. Release a pen, and it would "fall" toward the floor, since you and everything else are being boosted in the other direction. Einstein called this the *principle of equivalence*, about which he wrote:

> . . . we shall therefore assume the complete physical equivalence of a gravitational field and the corresponding acceleration of the reference frame. This assumption extends the principle of relativity to the case of uniformly accelerated motion of the reference frame.

In one of the most inspired steps ever taken in science, Einstein reasoned that since all gravitational fields vanish inside a free-falling observer's frame of reference, such as the astronaut inside his spacecraft in Earth's orbit, special relativity ought to apply to all measurements of distances and times made within that frame. And in a true leap of faith, he proposed that the two postulates of special relativity ought to apply even in cases where we compare the measurements of an observer in this frame with those of another individual far, far away, where the effects of the gravitational field are negligible. In this elegant and all-encompassing manner, special relativity and gravity were married; henceforth, the attraction of one object by another was to be described via its equivalence to an accelerated frame, and the laws of physics were to be written using the language of special relativity, though *generalized* to include relative accelerations as well as velocities between the various observers.

Having said this, however, we should also acknowledge that a crucial assumption has been made here. Almost a century has passed since Einstein completed his work, yet we still do not know whether the speed of light is in fact as universal as he claimed. Does light move at the same speed even in conditions of superstrong gravity? No experimental test thus far has even come close to answering this question.[2] Distances and times in special relativity are compared between different observers using the simple relation derived from the constancy of the speed of light. But if this capability does not extend to the strong-field realm, it is safe to say we are not yet certain about how to relate what is happening, for example, near a black hole[3] to what we see a great distance away.

One of Einstein's greatest strengths was his ability to conceptualize a physical principle without the encumbrance of heavy mathematical formalism. The reasoning that led to the principle of equivalence is perhaps his most enduring achievement fashioned in this manner. Once this basic idea is understood, however, one needs equations that tell us exactly how matter and energy create their gravitational influence—a code of nature. In other words, the principle of equivalence allows us to evaluate the impact of a known gravitational field on its surroundings, but to complete the task, we also need to know how to determine that field in the first place.

Newton's ideas concerning gravity work well for slowly moving objects, and just generally for systems in which the gravitational force is weak.[4] He found empirically that the gravitational force in this circumstance goes inversely as the square of the distance between the two objects. This prescription breaks down, however, when in addition to mass, energy is also present in the system. One of the best known equations in physics, $E = mc^2$, tells us that mass is equivalent to energy, so the latter must also produce gravity; it is not enough to determine the force from the mass alone.

There is also that annoying business about action-at-a-distance that the classical physicists used to talk about incessantly. Back then, forces between particles were expressed solely in terms of their position, as if each of them could "feel" the influence of the other simply by its presence somewhere in the system. But now that we have shattered the myth of absolute space, physical laws cannot be written without restricting how the effects of those forces are mediated through spacetime. In Newton's theory, Earth experiences the Sun's gravity instantaneously; if the Sun were to suddenly deviate from its current position, Earth would feel that motion concurrently. In

2. See chapter 4.
3. We will examine these enigmatic objects in chapter 8.
4. By "weak" we mean the field one encounters on Earth, or within the solar system, as opposed to the superstrong field curving the spacetime near a black hole.

special relativity, however, no influence can travel faster than the speed of light, so Einstein argued instead that Earth would feel that sudden motion only 500 seconds later, this interval being the length of time required for light to reach us from the Sun.

Toward the end of the twentieth century, an impressive demonstration that the effect of gravity itself travels at the speed of light—in the form of gravitational radiation—would produce a Nobel Prize for its discoverers.[5] Contrary to the classical view that gravity should be felt instantaneously everywhere in space, its influence must instead be carried from point to point.

Between 1907 and 1915, several physicists struggled to develop the methodology for calculating the attraction due to gravity, correctly taking into account the contribution from energy as well as mass, and the finite propagation speed of gravitational waves. Numerous papers appeared on this topic, but none of them seemed to solve the problem completely. Einstein himself had made several attempts at publishing the correct equations, only to retract them with each revision. By the end of 1915, this sequence of mistakes prompted him to write to a colleague, "That fellow Einstein suits his convenience. Every year he retracts what he wrote the year before."

But by then he had found the ideal person with whom to correspond— one of the greatest mathematicians of the twentieth century, or any century for that matter. Einstein's exchanges with David Hilbert (1862–1943) (fig. 3.1) proved to be the final inspiration, apparently for both of them. In November 1915 Einstein and Hilbert engaged in a rapid exchange of letters and drafts, clearly progressing toward the final solution, but at times also lapsing into angry claims of "nostrification." Hilbert's paper entitled "The Foundations of Physics" would ultimately bear the correct equations of general relativity and was actually submitted on November 20, five days earlier than Einstein's manuscript entitled "The Field Equations of Gravitation," but a careful scrutiny of the archives shows that the version published in March 1916 contained a modification apparently inspired by Einstein's own work.[6] However, Hilbert's paper contained some important additional contributions to relativity not found in Einstein's work. The two had shared ideas openly leading up to this point, and clearly each had influenced the other's thinking.

Ironically, the benefits Einstein derived from his communication with Hilbert were revealed indirectly in his letter of reconciliation written to his colleague on December 20, 1915:

---

5. This is one of the four "pillars" of general relativity discussed in chapter 4. For a technical reference on this subject, see also Damour (1987).
6. The details of this exchange may be found in Corry, Renn, and Stachel (1997).

FIGURE 3.1. The great mathematician David Hilbert, shown here in 1910 a few years before his work with Einstein.

There has been a certain resentment between us, the cause of which I do not want to analyze any further. I have fought against the feeling of bitterness associated with it, and with complete success. I again think of you with undiminished kindness and I ask you to attempt the same with me. It is objectively a pity if two guys that have somewhat liberated themselves from this shabby world are not giving pleasure to each other.

Hilbert's approach had been quite different from that of Einstein, but as powerful as his technique was,[7] he realized that he could still not overcome a serious problem—how to demonstrate that energy is conserved in Einstein's theory. In classical physics, even in special relativity, it is possible for an observer to determine the energy of his or her system, whose conservation is then confirmed by showing that energy flowing through a closed surface must be balanced by its corresponding change within the volume. In general relativity, however, the gravitational field itself carries energy, but the amount is different for different observers. So the classical concept of energy conservation actually does not work when sources of gravity are present.

7. As a mathematician, Hilbert was drawn to the more elegant and powerful formalism known as *Hamilton's variational principle* in deriving the field equations of general relativity.

But by 1915 Hilbert had assembled in Göttingen a team of individuals to help him make sense of general relativity, including the thirty-three-year-old Emmy Noether (1882–1935), whose mathematical brilliance he had recognized early (fig. 3.2). Already by that time she had been developing what theoretical physicists now consider to be one of their most important tools—a theorem that relates symmetries in nature to the conservation of certain relevant quantities. Referred to as *Noether's theorem*, this grand opus makes it possible to understand, for example, why a translational invariance implies the conservation of momentum. Newton taught us that space has no center, that we could shift our frame of reference without affecting the physical laws. He also taught us that an object's momentum changes only if a force is applied to it. Noether's work explains why these two Newtonian discoveries actually mean the same thing.

Noether's profound insight into the connection between symmetries and conservation laws permitted her to define a new representation of the total energy of a system containing a gravitational field, demonstrating that its value is constant in time when one includes the contributions from all space.[8]

In his paper Hilbert attributes this result, written without proof, to Emmy Noether, who published the complete theorem under her own name in 1918. In their book *Symmetry and the Beautiful Universe*,[9] the American physicists Leon M. Lederman and Christopher T. Hill argue that Noether's theorem is "certainly one of the most important mathematical theorems ever proved in guiding the development of modern physics, possibly on a par with the Pythagorean theorem." Fleeing Nazi Germany in 1933, Noether eventually resettled at Bryn Mawr College, in Pennsylvania, but survived only two more years there. Following her premature death in 1935, Einstein wrote an obituary for the *New York Times*, in which he said:

> . . . the most beautiful and satisfying experiences open to humankind are not derived from the outside, but are bound up with the development of the individual's own feeling, thinking and acting. The genuine artists, investigators and thinkers have always been persons of this kind. However inconspicuously the life of these individuals runs its course, none the less the fruits of their endeavors are the most valuable contributions which one generation can make to its successors.
>
> Within the past few days a distinguished mathematician, Professor Emmy Noether . . . died in her fifty-third year. In the judgment of the most competent living mathematicians, Fräulein Noether was the most signifi-

---

8. An excellent account of this work is given by Byers (1999).
9. See Lederman and Hill (2004).

FIGURE 3.2. Emmy Noether, around the time she joined Felix Klein and David Hilbert at the University of Göttingen, where they were working to further define Einstein's theory of general relativity.

cant creative mathematical genius thus far produced since the higher education of women began. . . . [S]he found in America up to the day of her death not only colleagues who esteemed her friendship but grateful pupils whose enthusiasm made her last years the happiest and perhaps the most fruitful of her entire career.[10]

At last, Einstein's code of nature—the equations that describe the gravitational field in general relativity—had been found. Now what? As anyone who has ever taken the time to study them will acknowledge, their simple, elegant appearance belies the fact that deep inside they possess a horrendous complexity. Chief among the problems is the fact that Einstein's code

10. See "Emmy Noether," by Professor Albert Einstein, published by the *New York Times*, May 5, 1935.

FIGURE 3.3. Karl Schwarzschild, a professor of physics at the University of Potsdam, produced the first complete solution—a highly simplified, time-independent spacetime surrounding a spherically symmetric mass—to Einstein's field equations of general relativity.

is actually not one equation, but six, all interwoven, such that no single one of them can be solved without simultaneously solving the rest.

Yet one cannot test the ideas of general relativity—the nature of spacetime, the principle of equivalence, et cetera—without tracking a real object and comparing its behavior to that predicted from the theory. But so difficult was the task of even formulating what could be predicted, that Einstein himself could not solve his own equations—at least not in a way that permitted tests of the theory in regions where the fabric of spacetime is affected the most.

Of course, during the development of general relativity, Einstein was able to extract certain predictions of his theory, based primarily on a reduction of his equations to a form consistent with Newtonian gravity. By doing away with most of the complexity, he could demonstrate that the coupling between space and time prevents a planet on an eccentric orbit from forming a closed ellipse as it winds around the Sun. As we shall see in chapter 4,

Einstein's explanation for the advance in Mercury's perihelion was a dramatic early confirmation of his theory. But successfully venturing beyond such limiting cases would prove to be something that only others would (or could) do.

How impressive was it, therefore, that the first significant step in the eventual unlocking of Einstein's code was taken within only a matter of months of its publication? A professor at Potsdam, Karl Schwarzschild (fig. 3.3) had volunteered for military service in 1914 and had been stationed on the Russian front when he received copies of Einstein's papers. Amazingly, he had the foresight, while struggling to survive, to invoke the highest degree of symmetry he could imagine in order to eliminate as many of Einstein's six equations as possible.

In so doing, Schwarzschild produced a description of gravity surrounding a single compact object, such as the Sun or Earth. Sadly, he contracted an illness soon after his discovery and died upon returning home. At the time his work was considered to be purely theoretical, with little application to reality. Today, however, Schwarzschild's name is rightfully associated with several distinguishing features of black holes, about which we will have much more to say in chapter 8.

4 :: FOUR PILLARS AND A PRAYER

But though Einstein's equations of general relativity had already produced an intriguing description of spacetime showing a clear departure from the simple action-at-a-distance in Newtonian theory, the changes were too radical for scientists to abandon their cherished beliefs and immediately adopt the new framework. One is reminded of the fact that extraordinary claims require extraordinary evidence. In the case of general relativity, the evidence would indeed be extraordinary, winning a Nobel Prize for some of the experimenters who developed the tests. It would take at least four major breakthroughs, however, spread over six decades, before the world's relativists would develop full confidence in the viability of Einstein's equations. And it would take almost that long, as we shall see in chapters 5 through 8, for them to crack Einstein's code and find meaningful solutions they could use to compare with nature.

Having said this, it is also true that the prolonged testing phase of Einstein's theory received a quick and auspicious beginning in the years following Einstein's and Hilbert's historic papers. England and Germany were at war during this time, but Sir Arthur Eddington (1882–1944) somehow kept up with the scientific progress unfolding on the continent. A Quaker by birth, Eddington (fig. 4.1) himself never fought in the war, excusing himself on religious grounds.

Learning about general relativity through documents smuggled into Cambridge, Eddington devised an experiment to test one of its principal tenets using the Sun—the most massive object near Earth. By this time he had already become a convert to Einstein's theory, which could explain a hitherto puzzling phenomenon associated with the solar system's inner planet.

Astronomers had known for a long time that Mercury's orbit is not quite elliptical; its path around the Sun does not close at exactly the same spot every time the planet goes around. The quasi-ellipse rotates slowly, but measurably so, a phenomenon that cannot be explained with Newtonian me-

FIGURE 4.1. Arthur Eddington's eclipse expedition in 1919 provided the first confirmation of the prediction that gravity bends the path of light when it passes near a massive star.

chanics in terms of the Sun's gravity alone. This force always points to the center and cannot drag the planet forward along its orbital path. However, the other planets also tug at Mercury. In 1845 Urbain Jean Joseph Le Verrier (1811–1877) calculated the precession of its orbit using all the known objects in the solar system and succeeded in accounting for all but 35 arcseconds per century of the measured value. The discrepancy was confirmed in 1882 by Simon Newcomb (1835–1909), who determined a rate of advance of 43 arcseconds per century.[1] Attempts at finding other interplanetary matter that could cause such a deviation from a pure elliptical orbit met with little success.

The answer would eventually be found in general relativity, which modifies the lengths and time intervals we measure. When applied to Mercury's orbit, these modifications result in a measured circumference at that radius that is not quite the same as Euclid's formula.[2] Thus, the orbit we see is not an exact ellipse, and the distortion predicted by Einstein's theory is precisely 43 arcseconds per century, in perfect agreement with the experimental value.

1. Simon Newcomb's analysis of Mercury's orbit is discussed in Carter and Carter (2006).
2. In flat (or Euclidean) space, the circumference is simply given as $2\pi r$, where r is the radius of the planet's orbit.

With the success of this "first" test, Eddington was understandably eager to examine the theory's more vulnerable aspects—such as the bending of light. The light reaching us from distant stars, Einstein argued, ought to bend around the Sun on its passage toward Earth. The principle of equivalence, upon which general relativity is built, correctly states that the effects of gravity are equivalent to a global acceleration of the underlying frame of reference. Standing inside an elevator, one cannot tell whether we and everything around us are being "pulled" down by Earth's gravity, or whether the elevator is being accelerated upward. The discovery by Galileo Galilei and Sir Isaac Newton that the gravitational force on an object is proportional to its inertial mass made this possible since it "explains" why the acceleration is the same for all matter.

This reasoning, however, has very little to say about particles with *zero* mass—such as the photon constituents of light. Perhaps the behavior of light could after all be the distinguishing factor between the effects of gravity and those of an accelerated frame. Suppose we grant that for some unknown reason the gravitational force on an object is proportional to its inertial mass. Then the equivalence between the effects of gravity and those of an accelerated frame could apply to all *massive* objects but not necessarily to the *massless* ones. In that case, a light path would appear bent in an accelerated frame, but not in a gravitational field, since a ray of light propagating horizontally across the elevator (as seen from outside) would appear to be curving downward to the upward-moving observer inside.

Unfortunately, the most massive object near Earth also happens to be the brightest, making the observation of stars near its rim impossible when one sees the Sun itself. Einstein's clever solution was to carry out this experiment during a total solar eclipse, when the Sun's direct rays are blocked from reaching an astronomer's photographic plate, thereby allowing the much fainter glow of distant stars to also appear on the image.

Eddington was convinced of the importance of such an observation, which would directly demonstrate whether or not light obeys the principle of equivalence. Detecting a bend in its path would constitute a stunning affirmation of Einstein's ideas and an irrevocable break with the classical worldview. It was an opportunity he could not pass up.

Eddington was a very careful and methodical worker, and he organized the expedition to observe the eclipse with great care. He even split his team up into two groups, one heading for the island of Príncipe on the west coast of Africa, the other to Sobral in Brazil, fearing that sending one team alone would run the risk of being clouded out during the eclipse. This was rather prescient on his part, because indeed his own expedition to West Africa would greet the much-awaited day of the experiment with a thunderstorm and persistent showers. The other team, in Sobral, was much more fortunate.

FIGURE 4.2. A half-tone reproduction of one of the negatives taken by Eddington's group using the four-inch lens at Sobral, in Brazil. The positions of several stars are indicated with bars. When compared to other photographs taken of the same region of the sky, it became apparent that those closest to the rim of the Sun appeared to have shifted slightly. (Image from Dyson, Eddington, and Davidson [1920].)

Shortly after the mission, on November 8, 1919, the *London Times* proclaimed in a headline: "The Revolution in Science: Einstein Versus Newton." Two days later the *New York Times* echoed with "Lights All Askew in the Heavens: Men of Science More or Less Agog Over Results of Eclipse Observations; Einstein Theory Triumphs." The commotion around the world was being fueled by the results of Eddington's observations, which seemed to confirm that light rays from distant stars are deflected by the Sun's gravity in the amount predicted by general relativity (see fig. 4.2).[3]

The excitement generated in November 1919, which turned Einstein into an overnight celebrity, was ignited by the realization that everything—not just massive particles—is affected by gravity. Massless or not, light follows a curved trajectory, just as a baseball does when it is thrown toward home plate from left field. The presence of a massive object, it seems, changes the spacetime around it to make everything follow a curved path.

---

3. A more recent historical examination of the images produced by Eddington and his team has shown that the raw data were actually inconclusive, though later observations did unquestionably confirm the result. It appears that Eddington was arbitrarily selective in choosing which results to use. See, e.g., Hentschel (2005).

Be they cricket balls, protons, or photons, they all accelerate toward the source of gravity at the same rate. Thus, the conclusion had to be that if the gravitational acceleration is identical for all things, it ought to be due to a modification of the spacetime through which everything is moving. This idea has now been verified experimentally numerous times, and with greater and greater precision as instruments have improved over the years (see fig. 4.3).

This early confirmation provided at least some motivation for physicists to explore the consequences of Einstein's theory, but progress was slow and often came when one generation of scientists gave way to the next. About a decade later, J. Robert Oppenheimer (1904–1967) and colleagues began to ask serious questions about what happens when a star approaches the end of its life (see chapter 8). One of their conclusions—that catastrophic collapse to a singular point is inevitable—was so distasteful to many, including Einstein himself, that this effort essentially dragged to an inconclusive end, at least until the 1960s.

It didn't help much that the next major confirmation of general relativity would not come for almost forty years after the first two. Unfortunately, Einstein died a few years earlier, so he never even became aware of the incredible impact this test would have on how scientists now view his theory. Though this third major test is sometimes referred to as an experiment to measure the weight of photons (i.e., light),[4] it actually has to do with an important prediction of general relativity—that clocks should run slowly in a gravitational field.

To understand how this comes about, let us consider the fact that the acceleration of a particle in the vicinity of a gravitating source is independent of its velocity. Viewed from a distant vantage point, that particle accelerates at a rate determined by its radius from the center of the object, regardless of how quickly it is moving. In special relativity, the distortions to time and space are entirely dependent upon the relative velocity between two different observers. General relativity, however, introduces additional distortions that constitute more than a mere subtlety. To be sure, the special relativistic length contraction and time dilation must still apply to any pair of reference frames, whether or not they are imbued with a gravitational field, but the acceleration does something new—it causes clocks to run more slowly in the presence of gravity. The reason for this effect can be understood with a careful consideration of the meaning of time and its dependence on change.

---

4. Indeed, even the experiments themselves wrote this in the title of their paper. See Pound and Rebka (1960).

FIGURE 4.3. The galaxy cluster Abell 2218 provides us with a modern version of the light-bending excitement generated by the 1919 expedition to the west coast of Africa. Abell 2218 lies approximately 2 billion light-years behind the constellation Draco. Deep Hubble images reveal many beautiful arcs surrounding its center. Most of these arcs are actually gravitationally focused images of a single galaxy, about five to ten times more distant (and behind) the cluster. The cluster is so massive that its enormous gravitational field deflects light rays passing through it, much as an optical lens bends light to form an image. This phenomenon, called *gravitational lensing*, magnifies, brightens, and distorts images from faraway objects. In this particular case (which is rare), the light is bent so strongly that the image is sheared out into an arc. In addition, the strength of the lens is such that light passing to either side of the cluster gets bent toward the observer and multiple images form. (Courtesy of A. Fruchter at the Space Telescope Science Institute, and NASA.)

In any given frame, time passes when something is changing. To measure a second on a clock, the hand must turn one-sixtieth of a complete cycle across the clock's face. Let us suppose that the clock is inside an elevator accelerating upward at a rate of two feet per second every second. Regardless of how fast the elevator is moving at this instant, during the next second on the clock, its speed will have increased by two feet per second. As long as the elevator is accelerating, there is no way to avoid the fact that its velocity relative to the outside world is different at the end of that elapsed second compared to its value at the beginning. But suppose we attempt to get around this change and instead let the clock run for half a second. This time the elevator speeds up by one foot per second, but again the velocities are different at the beginning and at the end. Unfortunately, no matter how much we shrink the interval of time, the starting and ending velocities are always different—that's the nature of acceleration, of course.

The same situation applies to the particle falling under the influence of gravity. If an observer moving with the particle were to measure time on its clock, he or she would cross into a frame of reference moving even faster relative to us than before, because while measuring time, time itself has moved on. Thus, according to special relativity, there should be an additional time dilation associated with this increase in speed. The effect is greatly magnified when the acceleration is so great that even a minute interval can bring the magnitude of the particle's velocity uncomfortably close to the speed of light. The ensuing time dilation can then appear to freeze the action completely inside the particle's frame. Thus, as long as we accept the fact that time intervals and distances are altered during a transformation from one frame to another (but always in such a way as to preserve the constancy of the speed of light), we must also accept the conclusion that the acceleration of one frame relative to another itself incurs an additional time dilation. Gravitational fields, therefore, slow down the passage of time as viewed from distant vantage points, and the retardation effect is greater the stronger the field.

Several other important consequences follow immediately from this effect, including the often used (and observationally powerful) concept of a *gravitational redshift*. The distance between the crests of a wave, called the *wavelength*, changes in inverse proportion to the frequency with which they are produced. The more frequently the oscillation peaks, the shorter the separation and hence the wavelength.

Now imagine our falling particle sending out pulses of light, which travel through the gravitational field out to large distances where we view this activity. Inside its reference frame, the crests of the wave are separated by a certain time interval. But according to us, who view these crests from far away, the interval is longer because of the dilation caused by the gravitational acceleration. So we infer a smaller pulse frequency (i.e., a smaller number of pulses per unit time) than that measured in the particle's frame, and we therefore see a longer wavelength than that inferred by the observer attached to the particle. Time slows down, the frequency decreases, and therefore the wavelength increases. This effect is known as a *redshift* because light of a certain color is shifted toward the red end of the spectrum where the wavelength is longer. The stronger the gravitational field is, the more significant this effect becomes.

By now several experiments have shown that this phenomenon is real (and measurable!). Back in 1960 two Harvard physicists Robert Pound and Glen Rebka (fig. 4.4) directed the 14,400-electron-volt gamma rays from radioactive iron on the ground floor of a 21.6-meter tower on campus toward the top, where similar iron nuclei were positioned to absorb them. This process is supposed to be reversible, in the sense that if an iron nucleus can emit a particular gamma ray, it ought to be able to absorb it or another like

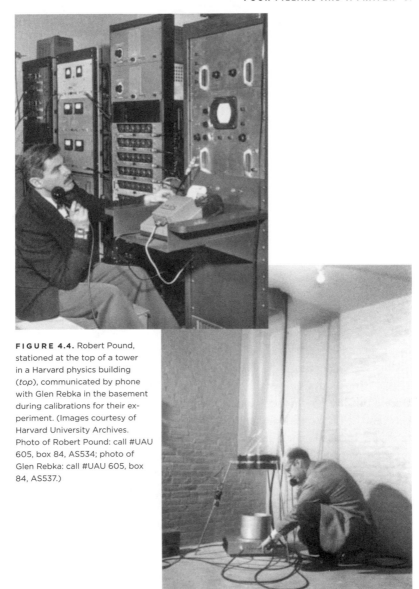

FIGURE 4.4. Robert Pound, stationed at the top of a tower in a Harvard physics building (*top*), communicated by phone with Glen Rebka in the basement during calibrations for their experiment. (Images courtesy of Harvard University Archives. Photo of Robert Pound: call #UAU 605, box 84, AS534; photo of Glen Rebka: call #UAU 605, box 84, AS537.)

it with very similar energy. Pound and Rebka reasoned that although Earth's gravity is weak, it is nonetheless not completely absent, and the difference in height between the base of the tower and its top should introduce a relative gravitational redshift. More specifically, gamma rays emitted on the ground should be redshifted by the time they reach the iron nuclei at the top, and this shift should be sufficient to quench their rate of reabsorption when they get there because of the mismatch in wavelengths.

Pound and Rebka were immediately excited by the realization that the measured absorption rate was indeed less efficient than normal. But when the iron nuclei at the top of the tower were made to move downward at just a particular speed, the gamma rays were then readily absorbed. The motion of the nuclei at the top was just enough to compensate for the gravitational redshift up the tower. This process works by analogy with the change in pitch (or frequency) that we sense when a train approaches and then recedes from us. Moving the absorber downward (toward the emitter on the ground) increased the frequency (i.e., "raised the pitch") of the gamma rays as seen in the frame of the nuclei, permitting them to be reabsorbed. This rather simple, yet elegant, experiment produced a fractional redshift within about 10 percent of the predicted value and is considered to be one of the major achievements of twentieth-century physics.

In 1976 the U.S. National Aeronautics and Space Administration (NASA) carried out its own version of this experiment using clocks instead of gamma rays, and with a much bigger distance between the two events. Called the Gravity Probe A (or GP-A for short), the NASA mission carried a hydrogen maser[5] clock to a height of 6,200 miles atop an expendable rocket; at this height, Earth's gravity is half as strong as it is at sea level. The probe slowed, stopped for an instant at the apex, and then started falling back. During this brief moment, the scientists could measure slight differences between the flight clock and another one kept on the ground. The gravitationally induced time dilation agreed with the value expected from the principle of equivalence to within 0.007 percent accuracy.

The Pound-Rebka experiment had a profound influence on the scientific community, rekindling widespread interest in general relativity. As we shall see in the next chapter, the general mood at the meeting on Gravitation and General Relativity in Poland just a few years later was one of confident anticipation that Einstein's theory would soon be brought into mainstream coexistence with the rest of physics.[6] The future Nobel Laureate Vitaly L. Ginzburg (fig. 5.9), who spoke on the experimental verification

---

5. The acronym "maser" (microwave amplification by the stimulated emission of radiation) was coined by the Nobel Laureate Charles Townes and his collaborators in the mid-twentieth century, after they successfully demonstrated the principle of microwave amplification in their laboratory.

6. Attendees at this inaugural meeting included the Nobel Laureate Paul Dirac (1902–1984) and future Nobel Laureate Richard Feynman (1918–1988) (fig. 5.7), who sought to "quantize" general relativity.

FIGURE 4.5. Joseph Taylor and Russell Hulse celebrate their Nobel Prize win in 1993, for their earlier discovery of a binary radio pulsar whose spin-down rate confirmed the release of gravitational radiation consistent with the prediction of general relativity. (Image courtesy of the Seeley G. Mudd Manuscript Library, Princeton University.)

of general relativity, was particularly enthused with the latest results and spread his excitement to the rest of the attendees. An especially bright and eager young man from New Zealand (fig. 5.10) sat transfixed at these proceedings; within a year of the meeting, he would be propelled into an encounter with history, a remarkable story that will unfold over the next few chapters.

A decade later,[7] a serendipitous discovery would initiate an entirely different test of Einstein's theory, which helped to permanently underscore its viability in our description of the physical world. Unlike the previous tests that were essentially confined to local effects near Earth, this one was truly astrophysical in character. It was so compelling, in fact, that its investigators would later win the Nobel Prize for their seminal work (fig. 4.5).

Pulsars are rapidly spinning neutron stars, as massive as the Sun but no bigger than a city. They are so condensed, that a teaspoon full of their material weighs as much as all of humanity combined. Drop a marshmallow onto their surface, and they release as much energy as a nuclear explosion on Earth. Not all neutron stars have a strong magnetic field, but those that

7. Essentially toward the end of relativity's golden age, which extended from 1960 to the mid-1970s. See chapters 6–10.

are magnetized produce beacons of radio waves in two oppositely directed cones that flash around like powerful searchlights (see fig. 4.6). If we happen to be aligned just right, the beam illuminates us here on Earth once (sometimes twice) upon every rotation, with a period often shorter than a hundredth of a second. Because they are so ponderous, and spinning freely, pulsars are among the best clocks in the universe, keeping time with exquisite precision.

Hundreds of pulsars had been discovered by 1974, but the one found with the 300-meter radio telescope at Arecibo, Puerto Rico, and named PSR 1913+16 by Joseph Taylor and his graduate student Russell Hulse was very special. From the behavior of its beacon signal, they deduced that this particular object was orbiting a similar companion at a distance only a few times that of the Moon from Earth. No other binary containing two pulsars had ever been seen before (see fig. 4.6). This system turned out to be different from the rest because its behavior could not be explained by simple Newtonian gravity, which permits orbits to last indefinitely. Two compact stars, each no bigger than a city and orbiting each other so closely, produce very noticeable (and measurable) relativistic effects.

We learned in chapter 3 that although there exist many ways to describe general relativity, there are essentially two principal differences between it and Newton's law of gravitation, when one reduces the theory to its irreducible elements. The first is that in Newton's description, the gravitational force is an action-at-a-distance, meaning that as long as a source of gravity (such as the Sun) is present somewhere in the system, another body (such as a planet) immediately experiences its force just by virtue of existing somewhere else, no matter how far away it is. But Einstein's theory, built on a foundation of special relativity, includes the condition that all gravitational influences travel at the speed of light. Thus, for example, if the Sun were to suddenly magically disappear, it would take Earth about eight minutes to respond to such a dramatic change, this being the time it takes light to travel between us and the Sun.

The second principal difference is that in Einstein's theory, energy is equivalent to mass, and therefore it too can produce a gravitational influence, or curvature, in the surrounding spacetime. Whereas the gravity of an assemblage of masses in Newton's theory is simply the sum of all their individual contributions, in Einstein's theory one must also take into account the interactions among the masses themselves in order to correctly calculate the full gravitational acceleration.[8]

---

8. This works a little like compound interest, in the sense that the overall return is augmented by the interest on the interest itself, in addition to that on the original deposit.

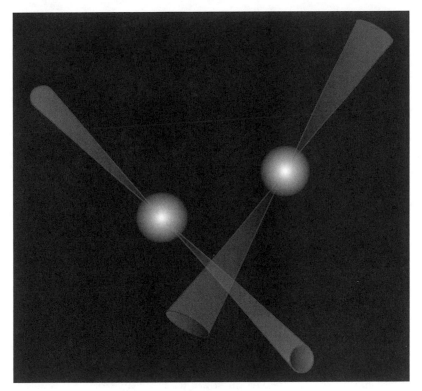

FIGURE 4.6. Two radio pulsars spin about their axes as they orbit each other. This tight configuration leads to the emission of gravitational waves, at a rate equal to that predicted by Einstein's field equations.

The Hulse-Taylor binary pulsar turned out to be a perfect laboratory for examining the gravitational consequences of the first effect. Because a body's gravitational influence propagates away from it at the speed of light, and because the surrounding spacetime therefore takes a finite amount of time to adjust to any changes in the system, a variable source of gravity ends up producing waves—gravitational waves—undulating away from it in all directions. This phenomenon is a unique feature of Einstein's theory and is one of its most important predictions.

But not every object can produce gravitational waves—even if it is variable. An important counterexample is a perfectly symmetric sphere, which may be pulsating radially. Due to the high degree of symmetry, the spacetime surrounding the sphere does not change at all, even as the source breathes in and out. In order to produce ripples in spacetime, the variable object must change aspect considerably as it vibrates or rotates. A rapidly spinning binary pulsar does exactly this, as the two compact stars plow through the surrounding spacetime.

The ensuing drag on the system does two things: first, it creates a wave train of fluctuations in spacetime, propagating away from the binary system; second, it slows down the stars, thereby decreasing their orbital energy—a process that saps their support against collapse and leads to a subsequent catastrophic merger.

The two neutron stars in the Hulse-Taylor system are so close to each other, and are orbiting at such speeds, that the shift in their elliptical trajectory due to the effects of relativity is four degrees per year. This is the same effect that astronomers had noticed earlier in Mercury's path around the Sun, only much, much bigger—about 33,000 times as large! This result by itself is already a useful confirmation of the first test of relativity, but the binary pulsar's importance rests primarily on what it can teach us about gravity waves.

The time between two beacon sweeps (0.05903 second) of the pulsar that we can see directly has proven to be extremely stable, its period increasing by less than 5 percent per 1 million years. Considered as a clock, its precision is therefore comparable with the best atomic clocks on Earth.

But even so, this period does vary, and it is the variation that teaches us about the behavior of this unusual system. As the pulsar orbits about its companion, the pulse period modulates by several tens of microseconds. This is the Doppler effect, similar to the change in pitch of a train whistle as it first approaches and then recedes from us. From the size of the modulation, we can tell how fast the pulsar is moving and how big its orbit is, and this is sufficient for us to identify its binary partner.

It was actually a more subtle variation, seen over many years, that produced the Nobel Prize–winning result. Contrary to what Newton's law would predict, the orbital period of the Hulse-Taylor pulsar is declining—the two neutron stars are rotating faster and faster about each other in an increasingly tight orbit. It turns out that the degradation in the orbit can be attributed entirely to the spacetime drag we described above, in which the stars produce gravity waves as they revolve about each other. After many years of observation, astronomers can now calculate that the orbital energy loss rate agrees with the prediction of general relativity to within about one half of a percent, another spectacular confirmation of Einstein's theory.

Three decades after the discovery of the Hulse-Taylor pulsar, astronomers are now preparing the stage to introduce gravitational wave detection into their panoply of observational tools. The excellent agreement between this binary's decay and the prediction of general relativity is viewed as indirect proof of the existence of gravitational waves, but the waves themselves have not yet been seen directly. The radiation emitted by the binary pulsar right now is too weak to be detectable on Earth with existing technology.

But by the end of the binary's evolution, when the neutron stars begin to touch, the final in-spiral will be quick, lasting only a fraction of a second, and the consequent churning of the spacetime surrounding the binary will produce outgoing waves of unprecedented power.

Of course, it will take the Hulse-Taylor pulsar billions of years to reach this final phase, so we have no hope of detecting the gravitational radiation from this particular binary. But there are many others like it across the universe, and current estimates suggest that this type of terminal convulsion occurs tens, perhaps hundreds, of times per year.

The development of an experiment to actually detect gravity waves on Earth began around the same time the binary pulsar was discovered and has continued onward to this day. Many different groups have contributed to this endeavor over the years, but the two principal teams are those led by Kip Thorne at Caltech and Rainer Weiss at MIT. The joint collaboration is now called the Laser Interferometer Gravitational-Wave Observatory, or LIGO for short, which consists of a sophisticated combination of lasers and mirrors spread across several kilometers of land. There are actually two such facilities in the United States, one in Louisiana, the other in Washington State; and a third (called VIRGO) is being built by a joint French/Italian collaboration near Pisa, Italy. Other teams around the world, from countries that include Japan and Australia, are developing their own additional facilities to add to the global network.

Though difficult to implement in practice, the idea behind how these instruments (may) detect gravitational radiation is rather simple to understand. Gravity waves are undulations in spacetime produced by highly variable mass distributions, such as a decaying binary system. Passage of such a wave past us would produce a temporary stretching or contraction of space, depending on whether we find ourselves in the "crest" or the "trough" of the wave. But this effect is so small that a highly sensitive technique must be used to measure the tiny change. After many attempts at coming up with the ideal method, the LIGO team finally settled on interferometry to do the work.

Light itself is wavelike, so two waves passing through each other can either produce a big crest (if the two individual crests lie on top of each other), a deep trough, or a complete cancellation if it happens that one wave's crest merges with the other wave's trough. The ensuing combination of light enhancements and cancellations produces what is known as an *interference pattern*. Laser interferometry is the technique of using laser light to produce such a pattern.

By judiciously arranging the mirrors and the direction of light rays within the device, one can hope that the passage of a gravity wave will stretch or squeeze the space between the end points of the interferometer just enough to cause the two laser beams to shift relative to each other by a fraction of a wavelength. The interference pattern should therefore shift too, since the location of the laser crests and troughs will vary as spacetime undulates around them. At the time of this writing, at least some of these facilities have started taking data, though only null detections have been reported thus far.

Few doubt that an experiment such as this will eventually work to some level, since we are certain now that general relativity is the correct description of gravity when its influence is relatively small. We do not yet know, however, how accurately the theory addresses nature when the conditions are extreme. And we should perhaps categorize this as the relativists' prayer—that Einstein's foundational assumptions and postulates are sound. If any of them should break down, much of what we will discuss in the next few chapters, particularly the structure of black holes in chapter 8, may be irrelevant in the end.

Indeed, many hope that by pushing the theory to its logical end, we may identify its strongest predictions. Ultimately, the implied properties of black holes will be compared with nature, retesting general relativity in regions where gravity is superstrong.

As we shall see, much of what we think we know about black holes rests on the critical assumption that special relativity is always correct in frames of reference free of gravity's influence. But suppose this frame itself is falling in a strong gravitational field. Could it be that an observer in this frame no longer measures the same speed of light as his strongly accelerated counterpart outside? That would do untold damage to the theory, for then we would have no obvious way of connecting intervals of spacetime from frame to frame (see fig. 7.2). Many believe such an outcome is unlikely, but until Einstein's theory is tested in regions where gravity is very strong, such as near a black hole, no one can be sure.

And so it was to be, on the heels of the Pound-Rebka experiment, that the world's leading relativists would soon set alight what we now call the golden age of relativity. We have progressed past that period in this chapter in order to include some of the more recent developments on the experimental side of things. But the turning point in this field can rightly be traced to the years immediately following that discovery of time dilation (or gravitational redshift), through the excitement generated at the meeting on Gravitation and General Relativity two years later in Poland, where the world's leading physicists and future stars gathered for a landmark event.

5 :: AN UNBREAKABLE CODE

As if to mark the transition in relativity from a period of stagnation to a new era of optimism, the Warsaw meeting of 1962 (fig. 5.1) opened on an almost apologetic note. Not quite sure what to expect, many of the leading physicists came with renewed interest. But they brought with them a feeling of uncertainty, having watched Einstein's theory of gravity develop only incrementally after it had splashed so spectacularly across the front pages of the world's newspapers over four decades earlier.

During the meeting, some would rationalize the small advances made as a precautionary—even necessary—introspection designed to deepen our mathematical understanding of Einstein's equations. But Leopold Infeld (1898–1968) (fig. 5.2), the chair of the organizing committee, seemed to capture the early mood of the crowd when, in his introductory remarks, he bemoaned the fact that general relativity had been allowed to lapse into a state of irrelevance:

> . . . already in 1936, when I was in contact with Einstein in Princeton, I observed that this interest [in general relativity] had almost completely lapsed. The number of physicists working in this field in Princeton could be counted on the fingers of one hand. I remember that very few of us met in the late Professor H. P. Robertson's room and then even those meetings ceased. We, who worked in this field, were looked upon rather askance by other physicists. . . . This situation remained almost unchanged up to Einstein's death. Relativity Theory was not very highly estimated in the West and frowned upon in the East.[1]

Infeld himself was one of those dedicated relativists who had been pounding their head on the wall in an attempt to crack the Einstein code in the decades prior to the 1960s. Who better to recognize the seemingly

1. See Infeld (1964).

FIGURE 5.1. Confidence in general relativity had undergone a resurgence toward the end of the 1950s, sparked by the groundbreaking work of Pound and Rebka in 1959–60. By 1962 the mood of relativists at the meeting on Gravitation and General Relativity was still somewhat apprehensive but noticeably upbeat. (Image courtesy of Andrzej Trautman and the photographer Marek Holzman.)

insurmountable difficulties faced by anyone foolhardy enough to proceed down this path? And yet he was one of the few who had at least made some progress.

Infeld's work with Einstein and Banesh Hoffmann[2] (1906–1986) (see fig. 5.10) attacked the problem indirectly, employing a technique that would eventually be aided by computers once they arrived on the scene. They studied the behavior of objects moving under the influence of their collective gravitational field, but because solving Einstein's equations is prohibitively difficult—if not impossible—for such a situation, they essentially employed a method of successive approximations.

What this means in practice is that they considered a binary system moving slowly, like Earth and the Sun, separated by a large distance compared to the size they would have if all of their mass were squeezed into a

---

2. Together, these three authors published a classic paper in 1938 entitled "Gravitational Equations and the Problem of Motion," which would be the foundation for much of Infeld's subsequent work in relativity.

FIGURE 5.2. Leopold Infeld (*left*) had been working with Einstein since the late 1930s trying to understand how objects move under the influence of each other's gravity. Hermann Bondi (*right*), a significant figure in relativity at King's College, London, was attacking the problem of identifying an object's gravitational radiation at large distances, where its effects are small and manageable with relatively simple mathematics. (Image courtesy of Andrzej Trautman and the photographer Marek Holzman.)

tiny volume, forming an event horizon. Guessing the trajectory followed by each object, they then fed this quasi-solution back into the equations, hoping that this process would produce an even more accurate solution for the motion at the end of their calculation. This (iterative) method is based on the supposition that every time one does this, the end result is an even better representation of spacetime than the one available previously.

In the end the work by Einstein, Infeld, and Hoffmann had some lasting influence over time, but not as much as they would have hoped for, because it simply did not tackle the problem of solving the equations of general relativity way down deep, where gravity is strong and the relativistic corrections are unavoidable. But their work did find a suitable application years later in the analysis of the binary pulsar's motion (see chapter 4 and figs. 4.5 and 4.6). The reason is simple. To have any hope of making progress with this approach, one must approximate the behavior of the system by analyzing its motion at large separation. But then Newtonian gravity is a good limiting representation of the motion, since relativistic effects in this regime are quite small. As noted in chapter 4, even Mercury, the closest planet to the Sun, has an orbit that deviates ever so slightly from the prediction of classical theory. To be sure, the perihelion advance (or its equivalent) measured in the case of the binary pulsar is much larger than that of Mercury, but the overall relativistic correction to its motion is still small enough that one can obtain sensible results with the approximations used in Infeld's approach.

This problem has resurfaced in more recent years because of its direct relevance to the detection of gravitational waves, for example, with LIGO and VIRGO (see chapter 4). A modernized version of Infeld's method is still used for these problems, though complemented heavily with computer simulations, which permit the investigator to follow the collapse and eventual merger of systems such as the Hulse-Taylor binary pulsar near the end of its evolution.

Even Hermann Bondi (1919–2005) (figs. 5.2 and 5.8), one of the most significant general relativists from that era, was unsuccessful in his attempts to probe the spacetime close to the source of gravity. Bondi had a remarkable career in many respects and is credited as a co-originator of the *steady-state* model of cosmology, in which the universe is forever expanding, though it maintains a constant density through the ongoing creation of matter to form new stars and galaxies. Until the cosmic microwave radiation was discovered, the steady-state cosmology was actually viewed more favorably than the competing big bang theory, but since only the latter can explain the pervasive low-temperature radiation, the steady-state model has now completely fallen by the wayside.

In general relativity, Bondi's principal contribution was in calculating the gravitational radiation produced by variable sources. Facing the same "insurmountable" hurdles as Infeld, however, he was forced to restrict his attention to the asymptotic limit (i.e., very far from the source of gravity), hoping that he could then trace his solutions back to the origin where the massive object was located.

Bondi's work depended critically on the use of approximations, and none of it could be transformed into an exact solution. He would learn that once gravitational radiation detaches from its source and propagates to large distances, most of the "memory" of the object that produced it is lost. Like Infeld, Bondi would see the development of computational algorithms push this kind of work well beyond his pioneering efforts. He and his co-workers made several meaningful contributions to relativity, particularly the nature of gravity waves at large distances, but their methods could not probe the spacetime much closer to the source itself.

However, Bondi's work did lead to the discovery of an important clue—an amazingly subtle property of light rays propagating through space-time—that would finally lead to the cracking of Einstein's code just a year later (chapter 7).

In 1962 Rainer (Ray) Sachs (fig. 5.3) was emerging as one of the stars in the new generation of physicists and mathematicians then entering the field of general relativity. He too shared the collective view that after the founding of general relativity at the beginning of the twentieth century, there followed a protracted period of minor advances, none of which really had any sustained impact. In his talk he characterized the situation as follows:

> Since 1916 we have had a slow, rather painful accumulation of minute technical improvements which have advanced our understanding of the mathematical content of this theory and the physics of gravity. I think that the attempt to continue obtaining such minute improvements constitutes a legitimate and fascinating part of mathematical physics. If something really exciting turns up, fine. . . . Of course, it may happen that all our rather sophisticated attempts will be swept into obsolescence by some simple, wholly new idea or experiment.[3]

Indeed, within only a year of his presentation at this meeting, the major breakthrough would come at the hands of Roy Kerr, in part from the inspiration provided by Sach's own "minute technical improvements."

As a member of Hermann Bondi's group at King's College, London, Sachs had quickly learned the techniques being used at that time to study gravitational radiation far from the object that produced it. What he noticed, however, was that Einstein's equations could be simplified considerably if the *shear* in the bundles of light moving along geodesics were to drop off quickly as one recedes from the source of gravity—a major, major clue. But what does all this mean?

A *geodesic* is the shortest distance between two points in spacetime. For

3. See Sachs (1964).

FIGURE 5.3. Ray Sachs, a former associate of Hermann Bondi, took solutions of Einstein's field equations with the simple asymptotic properties developed by the group at King's College, London, and examined their dependence on distance from the source. (Image courtesy of Andrzej Trautman and the photographer Marek Holzman.)

flat space—meaning the absence of any gravitational influence—the geodesic is just a straight line. On the surface of Earth, the pilot of a jet plane wishing to save time and fuel flying from New York to Paris actually does not fly along a straight line as seen from the ground, but instead veers off slightly to the left and then arches back to the right toward Paris after crossing the halfway point. This arched path is shorter than a "straight" line from New York to Paris because Earth's surface is a sphere, and a straight line on the sphere is actually curved along its circumference, making it longer than the arched trajectory. Light rays emitted anywhere in spacetime follow geodesics outward from their point of origin, because these are the most "direct" paths they can follow.

A group of geodesics is said to be *shear-free* when these light rays follow special paths that prevent distortions to an image. The image may shrink or get enlarged, but it may not be distorted in any other way (see fig. 5.4). First considered by Sachs, this property was later used, in their own attempt to solve Einstein's equations, by Ivor Robinson (fig. 5.5) and Andrzej Trautman (fig. 5.6), who wrote about it as follows:

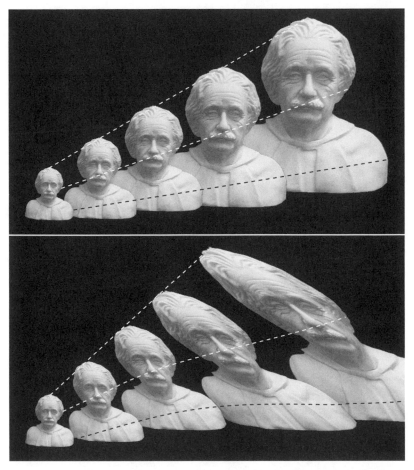

FIGURE 5.4. Einstein's field equations possess certain solutions with properties that permit some light rays to be shear-free. The physical meaning of this concept is illustrated in these two panels. In the upper image, the shape of Einstein's bust remains unchanged as we move along the null geodesics—essentially along a set of light rays in this particular spacetime—though its size may change and the image may even get rotated. In the lower panel, we see what can happen when the light rays in this spacetime are not shear-free, for then the shape gets distorted by the uneven spreading (or contraction) of the geodesics.

Think of the null geodesics as rays of light. Consider a small, plane, opaque object and a plane screen, some distance apart from the object. Suppose that the object and the screen are oriented so that they are orthogonal to the rays of light . . . and situated so that the shadow cast by the object can be observed on the screen. The [geodesics] are non-sheared if the shadow, as observed on the screen, is similar in shape to the object.[4]

4. See Robinson and Trautman (1964).

FIGURE 5.5. Together with Andrzej Trautman, Ivor Robinson (*shown here*) was exploring spacetime solutions of Einstein's field equations with special shear-free properties described in the text. (Image courtesy of Andrzej Trautman and the photographer Marek Holzman.)

FIGURE 5.6. Andrzej Trautman (*shown here*) and Ivor Robinson developed solutions of Einstein's field equations with special shear-free characteristics. (Image courtesy of Andrzej Trautman.)

Sachs could think of no physical reason why geodesics should behave in this way, but producing an asymptotically shear-free set of light rays certainly reduced Einstein's equations to a very manageable form. Robinson and Trautman went one step further. Why bother making the shear drop-off rapidly with distance? Why not just consider shear-free geodesics everywhere? They did find some solutions to Einstein's equations with this condition, but none of them acquired any direct relevance to real physical systems. They were seeking solutions corresponding to a variable source of gravity, again with the intention of exploring the gravitational radiation it produces; and by not going one or two steps further in the simplification of Einstein's equations, they were led in directions other than the one followed by Kerr just a few months later (see chapter 7).

Ray Sachs's "minute technical improvements" were to be his principal (and enduring) contributions to relativity. Only six years later, he would get disenchanted with this area of study and would change fields altogether. He became a biological physicist at Berkeley, where he spent the rest of his scientific career. Even after his subtle, though influential, contribution that lead to Kerr's dramatic breakthrough, it appears that he too succumbed to the ponderous drag that general relativity was exerting on most of its practitioners in the early 1960s.

So unbreakable was Einstein's code, that many shunned the issue of solving his equations altogether and opted instead to study the theory in the context of other active disciplines in science. The leading physicists of that day were understandably fascinated with quantum mechanics, an undeniably correct description of nature. The fact that the Pound-Rebka experiment had just affirmed the viability of general relativity as well caught the attention of those—such as Paul Dirac, Richard Feynman (fig. 5.7), and Peter Bergmann (1915–2002) (fig. 5.8)—who wondered how Einstein's classical theory could possibly be reconciled with the evident quantum character of the universe.

The challenge comes about specifically because of the way in which these two theories describe phenomena. Both disciplines use the same coordinates to identify position and time, and, perhaps surprisingly, even quantum mechanics does not attempt to alter the *continuous* character of spacetime. At least on this level, Einstein's theory and quantum mechanics agree.

But Einstein argued for the reality of a particle's properties and a precision in its location, regardless of whether anything else was interacting with it. This could be an observer using his or her ruler and clock to measure the particle's velocity, for example, or it could be another particle colliding with it.

Quantum mechanics holds that the particle's properties are defined only

FIGURE 5.7. At the 1962 meeting on Gravitation and General Relativity in Warsaw, Poland, the Nobel Laureate Paul Dirac (*left*) and future Nobel Laureate Richard Feynman (*right*) reminded the participants that, though general relativity was a classical theory, quantum effects could not forever be ignored. (Image courtesy of Andrzej Trautman and the photographer Marek Holzman.)

insofar as measurements can be made or an interaction can occur. In the absence of any "physical" contact with something else, the particle has no reference point, or standing, with the rest of the universe. Therefore, no one—not even an observer sitting on the particle itself—can know exactly where it is located or how fast it is moving, even if measurements are made, because all such determinations incur an error.

FIGURE 5.8. One of the leading (father) figures in relativity in the early 1960s was Peter Bergmann, shown here on the right, talking with Hermann Bondi. Bergmann's main effort at that time was focused on the quantization of gravity. (Image courtesy of Andrzej Trautman and the photographer Marek Holzman.)

Participants at the 1962 meeting on Gravitation and General Relativity in Warsaw could tell that general relativity was on the move, but no one knew yet where the next major advances would occur. Many of the seasoned veterans (see fig. 5.10) were hoping to uncover physically meaningful space-time solutions to Einstein's equations. The Schwarzschild metric, which had been around for more than forty years, was at best an approximation. Many even suggested that because real stars spin and their spin does not go away when they collapse, the effect of their rotation somehow prevents the formation of (what would later be called) a black hole. In other words, they argued that Schwarzschild's highly simplified solution could be used to predict the existence of singularities only because it ignored a crucial real effect in nature.

However, the newer entrants to the field, including Feynman and Dirac, were more interested in reminding everyone that quantum effects could not forever be ignored. In preparing for this meeting, Dirac himself took the bold approach of considering particles as extended objects, not points, and studying their behavior in a gravitational field along the lines pursued earlier by Infeld and his collaborators, as well as the Russian Vladimir A. Fock (1898–1974) (see fig. 5.10), who broadened the scope of their work by including extended objects—that is, not just gravitationally interacting point-masses, but masses spread out over finite regions of space.[5] The ideas presented by Dirac were early attempts at incorporating the "quantum nature" of particles into a description involving Einstein's equations.

Feynman went directly to the heart of the matter and simply pushed for a quantum theory of gravity. Many decades have passed now since those early attempts at "fixing" general relativity, but the truth of the matter is that to this day a self-consistent quantum theory of gravity is still only a dream.[6] Feynman himself became quickly discouraged and abandoned the effort. His attendance at the 1962 meeting on Gravitation and General Relativity would be his last.

But general relativity moved on very successfully from that meeting in Warsaw, hot on the heels of the Pound-Rebka experiment, and entered its golden age of discovery that lasted into the mid-1970s. The advances would come not from quantum mechanics, but rather from the more classical approach of solving Einstein's equations to obtain a description of spacetime associated with various distributions of matter and energy.

The attendees departed with elevated spirits, thanks in large measure to the efforts of Vitaly Ginzburg (fig. 5.9), who enthusiastically extolled the virtues of general relativity:

---

5. V. A. Fock was actually much better known for his contributions to quantum theory, where his name survives to this day, attached to often-used concepts and techniques, such as *Fock space* and the *Hartree-Fock method*.
6. To be sure, there exist several ongoing attempts to remedy this situation, such as *string theory*, which aims to unify everything in nature, though physicists are far from agreeing that the task is even close to completion.

FIGURE 5.9. At the 1962 meeting on Gravitation and General Relativity, the future Nobel Laureate Vitaly Ginzburg spoke on the experimental verification of general relativity. Ginzburg stressed the need to observe strong fields, noting quite emphatically the need to understand *rotational* effects in Einstein's equations. (Image courtesy of Andrzej Trautman and the photographer Marek Holzman.)

The widely-held opinion, entirely shared by the present author, is that general relativity theory unquestionably holds first place among all physical theories for its internal consistency and beauty.... Therefore, the present author (and, apparently, the majority of physicists) absolutely does not understand the tendency to build up non-Einsteinian gravitational theories; there is no evident physical foundation for doing so, not to speak of facts.[7]

Ginzburg shared the 2003 Nobel Prize in Physics for his fundamental contribution to our understanding of superconductivity, a phenomenon in which some materials carry currents freely, without any resistance, by virtue of a quantum effect that becomes important at very low temperatures. This work was actually carried out in the 1950s, so there was no one better to understand the importance and relevance of quantum mechanics at the time of the 1962 meeting on Gravitation and General Relativity in Warsaw.

7. See Ginzburg (1964).

FIGURE 5.10. In the front row of this audience for John Synge's description of slow-moving objects in the field of massive bodies at the 1962 meeting on Gravitation and General Relativity are (*left to right*) Leopold Infeld, Vladimir Fock, James Anderson, Ted Newman, Roger Penrose, and Banesh Hoffmann. The young lady to the right is Roza Michalska-Trautman, wife of Andrzej Trautman, who published several papers with Infeld on gravitational radiation. Among the seasoned relativists, a young Roy Kerr (seen between Fock and Anderson) sat attentively. (Image courtesy of Andrzej Trautman and the photographer Marek Holzman.)

Yet Ginzburg was an ardent supporter of Einstein's theory. He was convinced that general relativity was the correct description of nature, since it contains no variable parameters. And after all, every test that had been completed to that point fit its predictions remarkably well.

However, he did leave the attendees with one important admonition. He would not gloss over the fact that all the tests carried out by 1962 had examined the predictions of Einstein's theory in regions of weak gravity. One must push further, he beseeched the audience, much further, and pointed to the frame-dragging effects discovered by Josef Lense and Hans Thirring[8] in 1918 as a prediction of general relativity that might be accessible with satellite technology. As we shall see shortly, an important outcome of Einstein's description of gravity is that spinning masses force spacetime to

---

8. This effect in the context of spacetime surrounding a spinning object is described in chapter 7.

swirl around them in the same sense as their rotation, like a whirlpool of water orbiting around a drainpipe.

Ginzburg recognized the importance of strong gravitational fields and even suggested neutron stars as likely objects bound by strong gravity. At that time rapidly spinning neutron stars like those in the Hulse-Taylor system had not yet been discovered, though they were expected on theoretical grounds. But he was also keenly aware of the fact that any such object with strong gravity would also carry significant spin—it would have to, because every star we know of in the universe rotates, and if strong gravity is created when one of these objects collapses to an even tighter volume, its rotation rate can only increase, not disappear.

So toward the end of his presentation, Ginzburg emphasized the need to consider *rotational* effects in Einstein's equations. Near the front of the lecture hall, during most of the meeting, sat a young mathematical physicist from New Zealand who had just completed his doctoral thesis at Cambridge. Nestled among the seasoned veterans, Roy Kerr's attention was focused exclusively on the speakers, absorbing their every word. By the time he was ready to return to the United States, there would be no doubt left in his mind that he had found the right challenge.

6 :: ROY KERR

The flight on the military transport plane from Warsaw back to Dayton, Ohio, was long, tedious, and noisy, though uneventful. For Roy Kerr, however, there was precious little time to waste. The first meeting that summer, at the Department of Physics of the University of California, Santa Barbara, had not produced anything particularly new. But the July 1962 meeting on Gravitation and General Relativity in Poland had enthused all of its one hundred or so participants.

The time was ripe to delve deeply into the properties of Einstein's field equations, but there were other pressing matters to attend to as well. For one thing, Alfred Schild at the University of Texas at Austin had just founded the Center for Relativity, which in time would become one of the most influential institutes of its kind in the world, flourishing to this day. But it would take time for Schild to hire permanent staff members and, opting instead to recruit bright young scientists to bridge the effort over the first few years, he had offered Kerr a temporary position after the two had met in Santa Barbara that summer.

There was work with Josh Goldberg to finish off in the relativity group at Wright-Patterson Air Force Base, packing to do, and arrangements to be made moving a young family across the country. Somehow, everything fell into place, and by August 1962 Roy Kerr (fig. 6.1), his wife Joyce, and their young daughter, Susan, soon found themselves pulling up to the campus security gate at the University of Texas (UT) at Austin. As they waited for clearance, they reflected on the whirlwind journey of the previous five years—a transcontinental escapade that took them from idyllic Christchurch in New Zealand, to the ivy-covered walls of Cambridge University, and finally to the United States, where they were now hoping to make their permanent residence.

The rapid succession of events over the previous two years, first at Syracuse University and then at Wright-Patterson in Dayton, Ohio, had become

FIGURE 6.1. Roy Kerr, shown here in his apartment after joining the faculty at the University of Texas, Austin. (Image courtesy of Roy Kerr.)

a blur of new beginnings, premature endings, a stream of excitement, and the start of a young family. They remembered their ship out of England (the USS *United States*) docking in New York Harbor, and the immigration officer turning to Joyce as he exclaimed, "Oh, I see you have an abdominal tumor!" after inspecting her X-ray image.

Turning white, Joyce could muster only "What?" for a timid reply. The officer laughed and said, "No, you're pregnant! The fetus shows up on the X-ray." Though at first excited and amused at this unorthodox announcement of the pregnancy, the Kerrs quickly realized its implications.

They had decided upon leaving England that this would be a good time to start a family. Roy Kerr's career seemed to be on the move, and employment prospects in the United States were bright. Needing a visa to emigrate to their newly adopted country, they visited the American embassy, where, as part of the formal application procedure, they were given a physical examination. In those days this process included an X-ray, and the images were to be incorporated into the formal set of documents carried by the travelers.

Now, as they left Manhattan and headed upstate toward Syracuse, the possible consequences of those X-rays on the health of their unborn child became all too clear.

Susan was born in June 1959, a happy, vivacious baby who brought pride and happiness to the Kerr family. But she did not develop quickly as she grew—noticeably slower than other children her age. A year later Joyce returned to New Zealand with Susan to visit family and friends, who advised her to seek medical counsel. Susan, it turned out, was microcephalic, meaning that the child had an abnormally small head and an underdeveloped brain. The X-rays had done their damage. By the age of six, her convulsions were so bad that she needed placement in an institution. She died a year later, unable to recover from the permanent injuries she had suffered before birth.

The UT security gate opened, and as Roy Kerr edged the family car forward toward the central administration building, his thoughts retraced his days in Gore, a small farming community on New Zealand's south island. His parents had divorced before World War II, and when his father left for the military conflict in Europe, the young Roy (fig. 6.2) was sent to live with friends of the family who owned a farm near that town.

He remembered following his father to Christchurch at the end of the war, where his mathematical talent started to emerge, first through menial activities and later in more formal classroom settings. For a while he was employed counting rubber bands quickly and over long periods, in a small private factory, which also happened to be where he rested his head at night. In the late 1940s, times were harsh and resources scarce, and Roy's job was to package rubber bands for use in jam-jar covers. By spreading these out on a table and looking for patterns, he could count them at twice the speed (roughly six hundred per hour) of other workers.

The young Roy began his formal education in earnest upon enrolling at St. Andrew's College, a high school on the periphery of the city (fig. 6.3). At that time St. Andrew's did not have a mathematics teacher, so in his last year there, Roy was sent for additional tutoring by a coach whose task it was to train remedial students; in Roy's case, however, the coach was to prepare him for advanced standing at the university.

That year Roy would come to the attention of New Zealand's mathematical community when, still a student at St. Andrew's, he took the scholarship exams to attend university (fig. 6.4) (he later chose to stay in Christchurch and attend the University of Canterbury). Back then the mathematics exams consisted of two papers and were scored on a basis of 600 points. The top students each year received a scholarship to attend the university of their choice in New Zealand. Kerr got 298 in the mathematics portion of the test and finished thirtieth on the list overall, a rather sad and unexpected result. However, this disappointing outcome could be entirely explained by the fact that he had turned up in the afternoon for an exam given that morning, therefore receiving a score of zero. This oversight notwithstanding, the University of Canterbury recognized Roy's talents and permitted him to attend advanced courses.

In his first year at the University of Canterbury, where decades earlier the Nobel Laureate Ernest Rutherford (1871–1937) himself had spent most of his days before leaving for England, Kerr attended lectures for third-year

courses, though regulations allowed him to take only the second-level examinations for credit. Even the best students had never before been permitted to skip the first-year classes; Kerr was the first to do so, proceeding directly to the third level. He sat in on the master's level courses the following year. It was quite evident to all concerned that the educational resources in Christchurch, which he had completely exhausted before the second year had ended, were no match for Roy's abilities.

Midway through that second year at the University of Canterbury, the dean of science asked Roy why he hadn't applied for a postgraduate scholarship to go abroad. He could have gone to the University of Cambridge then and received a more thorough undergraduate education before beginning graduate school. But his instructors had forgotten to tell him about this opportunity, and he was forced to remain in Christchurch for an additional *three* years, essentially biding his time waiting to complete his degree before embarking on a trip to England.

As a result, Roy's undergraduate career would not be entirely dedicated to mathematics and science. A warm, engaging fellow to begin with, he

FIGURE 6.3. Roy Kerr began his formal education in earnest as a teen at St. Andrew's College in Christchurch. Today its well-manicured grounds provide a peaceful setting for contemplation and study.

found many opportunities to relieve his boredom, playing billiards and learning the game of golf (fig. 6.5). At the urging of his father, he even took up boxing, a horrifying prospect in retrospect. In 1952 he represented the University of Canterbury (or Canterbury College, as it was then known) as a light welterweight in boxing at the Easter Tournament. By his own reckoning, he was hardly a contender, easily getting knocked out on several occasions. William Sawyer, one of Canterbury's outstanding lecturers at that time, expressed alarm and dismay over Roy's pugilism, on the grounds that he didn't want the best brain he'd encountered in a student scrambled by a well-thrown punch. History will judge whether Roy came to any lasting harm over it.

In 1955 Kerr received a master of science (M.Sc.) degree with first-class honors from the University of Canterbury (fig. 6.6), and headed off to Cambridge with a Sir Arthur Sims Empire Scholarship. He boarded a ship with two other New Zealanders, also going for advanced study at Cambridge (fig. 6.7), and crossed the Pacific Ocean, headed for the Panama Canal, and then continued on across the Atlantic. A month later their ship finally pulled into port at Southampton, just a short drive from the university that would be their home for the next three years.

FIGURE 6.4. Roy Kerr (at sixteen) with his sister, Natalie, at the racetrack in Wellington, following a family reunion after years of separation and disruption due to World War II. (Image courtesy of Roy Kerr.)

FIGURE 6.5. Roy Kerr breezed through the undergraduate program at the University of Canterbury in Christchurch, affording him significant leisure time while waiting for his departure to the University of Cambridge in England. (Image courtesy of Roy Kerr.)

All students arriving in Cambridge from New Zealand enrolled for an undergraduate degree and would sometimes continue on to graduate study if they showed exceptional promise. It surprised no one that Roy Kerr, affiliated with Trinity College, was to be the first New Zealander to enroll directly into the doctoral (Ph.D.) program, his reputation having preceded him to the historic Cambridge campus.

FIGURE 6.6. In 1955 Roy Kerr (*second from the left*) graduated from the University of Canterbury and is shown here with colleagues and friends (including future physics professor John Elder immediately to his left) on the original campus in Christchurch. The inset shows Roy Kerr in 2007 at the same location, only a few steps from the Rutherford Museum, in what is now the Christchurch Arts Center. (Image courtesy of Roy Kerr.)

FIGURE 6.7. Roy Kerr (*left*) with fellow New Zealanders Graham Fraser (*middle*) and Mark Barber (*right*) at Cambridge in 1957. All three began their graduate studies at the same time, after spending a month together on the ship to England. (Image courtesy of Roy Kerr.)

Such was his promise as a mathematician, that Britain's greatest algebraist, Professor Philip Hall (1904–1982), accepted Kerr as a doctoral student to complete a thesis under his supervision. The son of unwed parents, Hall (fig. 6.8) rose through the academic ranks in England to become one of the country's most highly visible and respected mathematicians. Since his death, he has been described as one of the greatest mathematicians of the century, having contributed significantly to the recognition of group theory's importance to the physical sciences. This is a rather esoteric topic in pure mathematics, having to do with how elements with a defined commonality that links them together in a family behave with each other according to a prescribed set of rules.

With his "nineteenth-century" preparation in mathematics at the University of Canterbury, Kerr hardly knew what a group was, let alone have any aspirations of making a significant contribution to this abstract field. But confident in Kerr's exceptional ability, Hall assigned his student three test problems.

The first involved the *axiom of choice*, a component in the important branch of mathematics dealing with *sets* of elements, be they numbers, geometric shapes, or different colored cats. The axiom of choice is an assumption in this theory that makes it possible to form sets by choosing an element simultaneously from each member of an infinite collection of other sets, even when no algorithm exists for making the selection. This may sound obscure and poorly motivated, but it is actually an essential ingredient of many proofs in mathematics.

Dispatching this problem with ease, Kerr felt his spirits rise. He solved the next problem with equal speed and efficiency. But he made no progress at all on the third and final test. Admitting defeat, he sheepishly sought advice from his distinguished thesis adviser, who would proceed to give him a one-hour lecture on the subject. Pure mathematics was not for him, Kerr decided, and instead turned to mathematical physics, where the physical concepts are just as important to the problem as are the mathematical techniques themselves.

Years later Kerr would discover that Hall had assigned him one of the "great unsolved problems" of algebra—to prove the so-called general *Burnside conjecture*, having to do with properties of finite groups—as the third, innocent little test. A decade later it was actually shown by way of a counterexample that the conjecture was *false*. Hall's comments were always significant, but he was a modest man of few words. He has been described as gentle, kind, the soul of integrity, and amused. Kerr wondered how history might have changed had Hall revealed his mischievous little prank. But given where general relativity would soon take him, Kerr had no regrets.

FIGURE 6.8. Cambridge's Professor Philip Hall, one of the century's greatest mathematicians, took Roy Kerr as his student upon the latter's arrival on campus in 1955. (Image courtesy of the British Royal Society.)

Over the next year, Kerr took courses on quantum field theory from the Nobel Laureate Paul Dirac, and particle physics from the future Nobel Laureate Abdus Salam (1926–1996), then also a professor in Cambridge. But ironically, it would be neither of these distinguished scholars who would have the most profound influence on his thesis and subsequent research career. Roy soon met another graduate student, John Moffat (fig. 6.9), who introduced him to general relativity and inspired an interest in the motion of arbitrary spinning particles in spacetime. The more Kerr delved into the literature on this subject, the more he realized how severely limited all previous work on this topic had been. He had found an area rich in opportunity, for which his mathematical talents could greatly contribute significant new ideas and techniques suitable for problem solving.

Kerr found Moffat to be a very interesting individual, and the two friends talked often about a wide range of topics. Moffat actually started out his professional life as a struggling artist, but after a period of no income and starvation in Paris, he had moved to Copenhagen, where he learned mathematics and physics. Within a year he was working in general relativity and unified field theory—an all-encompassing description of nature that purports to unify the various known forces. His particular interest was in the unification of gravity and electromagnetism.

At age twenty he wrote a letter to Albert Einstein telling him that he was working on one of the great physicist's theories. Einstein wrote back in 1953, but the letter was in German, so Moffat ran down to his barbershop in

FIGURE 6.9. At Cambridge Roy Kerr met fellow graduate student John Moffat, who introduced him to general relativity. At that time John was working on unified field theories (attempting to combine general relativity with electromagnetism) for his thesis. (Image courtesy of John Moffat.)

Copenhagen to seek a translation. Many subsequent letters followed back and forth, particularly on the subject of quantum mechanics and its relevance to nature, which Einstein never found fully satisfying. But Moffat was not destined to follow in Einstein's footsteps, becoming instead what one might describe as an apostate of general relativity in later years.

In the early 1990s, Moffat published a controversial paper[1] proposing that, contrary to the second postulate of special relativity, the speed of light is in fact not constant. Without this constraint, general relativity itself collapses, at least in its most "general" form, so Moffat's proposal is a radical renunciation of Einstein's work. The speed of light may be constant now, he argues, but that does not mean it had to be constant all the way back to the big bang. By breaking Einstein's rule that the speed of light has never changed, he is proposing a possible resolution of the dark energy problem, whose existence is inferred from the apparent current acceleration of the universe. If the speed of light were greater in the past, the argument goes, then what we think is a present acceleration is merely an artifact of the shorter distance light is covering now, during a given time interval, compared with the past.

But in the late 1950s, Moffat was fully immersed in Einstein's theory, providing the initial inspiration for Kerr's interest in a field that has occupied the rest of his life. In his last year at Cambridge, Kerr produced an exhaustive thesis on particle motion in general relativity, which subsequently spawned several influential papers. Before leaving, he learned of the extensive community of relativists around the world, including Hermann Bon-

1. The idea that light might have a variable speed resurfaced later, e.g., in Joao Magueijo's (2003) book, but Moffat (1993) appears to have been the first to propose this concept.

di's group at King's College, London (see figs. 5.2 and 5.8), whom he would visit on several occasions as his graduate studies were coming to an end.

Kerr's visits to London proved to be critical for his later seminal work in Texas, for it was here that he heard an important lecture delivered by Felix Pirani, a collaborator of Bondi's on their research into gravitational radiation. Pirani had become interested in a novel approach used to analyze Einstein's equations, pioneered by Alexei Zinovievich Petrov (1910–1972), a Russian mathematician studying all the possible algebraic symmetries of equations with multiple components.

Anyone who has looked at Einstein's field equations can appreciate how complex they appear to be at first sight. The concept behind his expressions (of which there are four—coupling the four dimensions of space and time) is rather straightforward to say in words: they describe how the curvature of spacetime is altered by the presence of matter (or energy) within it. But for reasons we will explore at greater depth in subsequent chapters, Einstein's equations have an intimidating number of terms in them when written out in full.

Petrov provided the first important step that would eventually lead to Kerr's remarkable breakthrough several years later. The Petrov classification of the algebraic symmetries of Einstein's equations that Kerr heard about during that lecture showed that under certain circumstances (related to the shear-free conditions discussed in chapter 5; see, in particular, fig. 5.4), a breathtaking simplification would reduce the number of terms in these expressions to the point where one could begin to tinker with possible solutions.

It was also around this time that Peter Bergmann visited Cambridge and, after meeting Kerr there, immediately offered him a postdoctoral job at Syracuse University, where he had built a strong group working in general relativity. This is where the young Kerr family was heading later that year when they heard about the ill-fated X-rays from the immigration officer in New York.

Kerr arrived in upstate New York with enthusiasm and energy for a vigorous period of research on relativistic particle motion (fig. 6.10), but his research received scant little interest from the group already there. Bergmann and his followers had been working for some time on merging general relativity with quantum mechanics (a feat that to this day has eluded the many people who have tried it), and that's the direction they appeared to prefer.

It took a few months for Kerr to realize that the technique they were using in Syracuse was pointless and simply would not work. The Bergmann group abandoned its approach to quantization soon thereafter, and the technique has never reappeared. Barely a year after joining the group in

FIGURE 6.10. After meeting Peter Bergmann on one of his visits to Cambridge, Roy Kerr was recruited to his group and moved to Syracuse University in late 1958. (Image courtesy of Roy Kerr.)

Syracuse, Kerr now saw little point to staying there any longer. Fortunately, Josh Goldberg (fig. 6.11) had just started his own relativity group at Wright-Patterson Air Force Base in Dayton, Ohio, with the sponsorship of the U.S. Air Force, which had set up aeronautical research laboratories for fundamental research. Offered a more senior position there, Kerr was happy to accept Goldberg's invitation to join him.

As we saw in chapter 5, relativists around the world were by this time concentrating on one of three principal problems: First, how to quantize general relativity? The early participants in this effort were Paul Dirac at Cambridge and Richard Feynman at Caltech and, as just noted, Bergmann and his group in Syracuse. Many others would later try their hand in this endeavor, including John Wheeler (1911–2008) (see fig. 8.4), who had spent the past several decades thinking about a marriage between quantum mechanics and general relativity; Roger Penrose (fig. 7.4); and Stephen Hawking, who thus far appears to have been the only one to make any significant progress in this direction. The process named after him—*Hawking radiation*—is a direct consequence of a quantum mechanical effect near the (classical) event horizon of a black hole. This phenomenon will be discussed later in the book.

Second, how to describe the motion of one body under the influence of another—that is, how to understand situations such as two stars orbiting each other at very short distances where general relativistic effects become important? The Polish group led by Leopold Infeld, often collaborating with Einstein, was certainly at the forefront of this effort, but Kerr himself had just made significant contributions to this area during his doctoral studies in Cambridge.

FIGURE 6.11. Josh Goldberg, shown here at the Kerr fest held in Roy Kerr's honor on the occasion of his seventieth birthday in Christchurch, New Zealand. (Image courtesy of Roy Kerr.)

Third, how can one unify the various forces of nature, including gravity? One may wonder why this is even an issue, but physicists feel that an aesthetically appealing description of the physical world is more than just an issue of beauty—often one finds meaning in simplicity and elegance, and a unified description of the physical laws is more likely to be correct than a disjointed one. John Moffat followed this path during his thesis in Cambridge. And Einstein himself spent many years on the pursuit of unification at the Institute for Advanced Study in Princeton, New Jersey, after moving to the United States.

In the background were the persistent, though generally unsuccessful, attempts at finding new solutions to Einstein's equations of general relativity with relevance to the physical world. After forty years of frustration, the relativity community was hungry for meaningful advances beyond Schwarzschild's seminal, though oversimplified, description of spacetime surrounding a spherically symmetric, static body. Real objects simply do not look like that in nature and, to borrow from Ginzburg's admonition at the 1962 meeting on Gravitation and General Relativity in Warsaw (fig. 5.9), one *must* understand rotational effects in Einstein's equations in order to finally test the correctness of this beautiful (though enigmatic) theory beyond any lingering doubt.

It would be during his two years at the U.S. Air Force Base in Dayton, Ohio, that Kerr would finally come around to confronting this problem head-on. In 1960, shortly after the start of his collaboration with Goldberg, Kerr started to pay serious attention to the Petrov classification. Unable to retrieve the Russian mathematician's original papers on this subject, they re-derived his results themselves—from scratch.

Soon it was time for the world's relativity community to gather for the 1962 meeting on Gravitation and General Relativity in Warsaw (chapter 5). This was to be a pivotal event, providing Kerr with a clear demonstration of

FIGURE 6.12. In 1962 Alfred Schild (*middle, back row*) founded the Center for Relativity at the University of Texas, Austin. By 1968 the faculty in the center had been reassigned to other departments. Schild is shown here with some members of the physics department. (Image courtesy of Roy Kerr.)

where things stood in general relativity. Upon his return from Europe, he felt empowered to forge ahead and tackle problems of significance with all the skills he could bring to the table.

And now that Kerr and his family had finally arrived on the UT campus, he looked forward to meeting up with Alfred Schild again (fig. 6.12). Alfred had made quite an impression on the young Kerr in Santa Barbara, and the two would become close collaborators and friends for years thereafter. But little did Kerr know that day what was about to happen over the next six months. The events leading up to the summer of 1963 would change the course of general relativity forever and power a surge in research on the physics of high-energy sources that has brought us to where we find ourselves today.

7 :: THE KERR SOLUTION

In retrospect, luck played a part. How could it not? Einstein's field equations are so complicated that without some assumption or strategy of attack, there is simply no way to proceed toward a solution. Schwarzschild's approach was to adopt the highest degree of symmetry he could think of and simplify the equations to the point where he only had to worry about the radial coordinate and time.

Nowadays, relativists can use both analytical means and numerical approaches on a computer to calculate the spacetime structure. Indeed, several interesting discoveries since the 1960s have been made from grid-based simulations such as this.[1] But in the absence of proper interpretation, numerical calculations may end up providing no more than a jumble of numbers—and it is the interpretation that physicists seek, for one cannot come away with a deep understanding of what is actually happening in a system without a visceral sense of cause and effect.

Settling into his new office on the ground floor of the Center for Relativity, Kerr set about identifying interesting problems to tackle during his one-year visit at UT. Though motivated to select an appropriate strategy to use in solving Einstein's equations, he at first refrained from this kind of work because others, including (Ezra) Ted Newman (fig. 8.1), had made it clear they were hot on the trail and suggested there was no point in entering the race late. A new colleague, Alan Thompson, soon arrived from England and moved into a nearby office. Thompson would be an important catalyst in Kerr's research in the coming months.

Based on his work over the previous two years, Kerr had already decided that the only sensible approach to solving Einstein's equations would be to consider spacetimes that contain shear-free geodesics—light rays that

---

1. Examples abound, but I mention here just a couple of representative papers: Price (1972); Shapiro and Teukolsky (1991).

do not shear (or distort) an image as they propagate (see fig. 5.4). Schwarzschild didn't know it at the time, but his own solution had this kind of geometry. It was therefore quite reasonable for Kerr to guess that a more generalized solution for a spinning object should also possess this special property.

As such, Kerr's shock and disappointment were quite evident when Thompson walked into his office a few months later, holding a paper from Newman and his colleagues proving that such solutions do not exist. Here was an experienced relativist (Newman), who had been tackling this problem head-on, and who had come to the conclusion that no shear-free spacetimes could be realized when the source of gravity has properties like those we observe in nature.

Kerr felt he had missed an incredible opportunity, but though he was crestfallen, he suspected there might be something wrong with Newman's analysis. He would find out later that two other colleagues, Ivor Robinson (fig. 5.5) and Andrzej Trautman (fig. 5.6), had similar misgivings but had not yet formulated a mathematically consistent objection to Newman's argument.

Kerr took Newman's paper and retired to his office, flipping through the pages. He had always been an impatient reader; it was against his nature to start with the introduction, plow through the main body of the text, and suffer through the conclusions. He browsed through the report until he saw something that caught his eye—an equation that clearly appeared to be important since the sum of terms on one side did not add up to be exactly zero on the other. It looked odd. He quickly determined that the equation was wrong. What they should have gotten was $0 = 0$, a pointless result, but one that clearly was not showing up in their calculation. Rushing to Thompson's office, he belted out: "They're wrong! They're wrong, and I can prove it."

The next few weeks turned into a furious cocktail of adrenaline, trancelike bouts of distraction, and the smoke from seventy cigarettes per day. There was little sleep. There were no guidelines to follow, no precedent to lean on. Kerr had been waiting impatiently for those working on the problem to complete their search, only now it was clear that he alone could actually push the problem through to a satisfactory conclusion.

Untethered, he drove himself mercilessly, knowing that the prize was within his reach. He was satisfied now that Newman had exhausted his pursuit and would not find his mistake easily. He was also fairly certain that few others even understood the issues, let alone had the ability to easily judge whether the Newman result was correct. They would have great difficulty figuring out a way of fixing the error in order to pursue their own attempts at solving Einstein's equations. Years later Kerr would confirm this suspicion by learning that only his colleagues Robinson and Trautman might have cracked the code themselves if given sufficient time.

It had been known for decades that Einstein's equations might become more manageable with the "proper" choice of coordinates to represent space and time. It is not always obvious that one wants to write these expressions in terms of simple distances that we measure with a ruler and time with a clock. But even a judicious choice of coordinates, along with the shear-free approach, still left Kerr with a set of equations that looked like they would not be solved easily, even though they were far more manageable than the complete set offered by Einstein.

There were too many terms left, and Kerr decided to follow Schwarzschild's lead and resort to symmetry for additional simplification. The goal was to eliminate as many variables as possible and seek a solution that remains constant in time. Nowadays, many of us are quite familiar with modern attempts by physicists to detect gravitational waves produced by highly variable cosmic sources. These systems produce gravitational radiation precisely because their spacetime is not constant, and analyzing them is a much more difficult problem, even more challenging than finding the solutions Kerr was seeking in the early 1960s.

In fact, it was gravitational radiation in spacetimes with shear-free rays that Robinson and Trautman were pursuing. They obtained some simple solutions, but none of them appeared to have any relevance to the physical world because once a source of gravity produces the radiation, those gravity waves detach from the objects that produced them and wander about the universe unfettered. With modern computers, one can predict with a fair degree of certainty what those waves look like—for example, when a neutron star wobbles or two black holes merge. But back then one had to rely on highly simplified situations to solve the equations, and it was difficult, if not impossible, to say anything specific about the source that produced the radiation by analyzing the properties of the gravitational waves themselves.

Had Robinson and Trautman looked at the time-independent form of Einstein's field equations using the shear-free condition, it is quite possible they would have found solutions with relevance to physical sources. They might even have come across what we now know as the *Kerr metric*. But fate would lead them in a different direction.

The combination of shear-free conditions, a choice of coordinates that incorporated rotational symmetry from the outset, and no dependence on time were the secret ingredients that within a matter of weeks would permit Kerr to write down equations that he could easily solve (see fig. 7.1). He knew

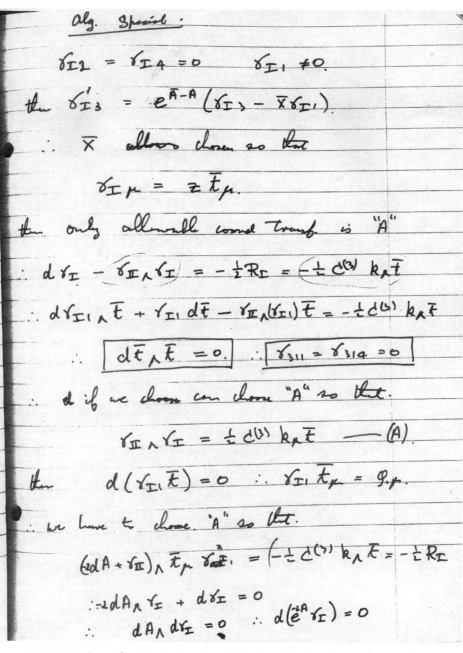

FIGURE 7.1. A page from Kerr's notes where he introduced the shear-free condition (see figure 5.4) to simplify Einstein's equations. (Kerr's notebook containing this page is held in the Alexander Turnbull Archives of the National Library of New Zealand, Wellington.)

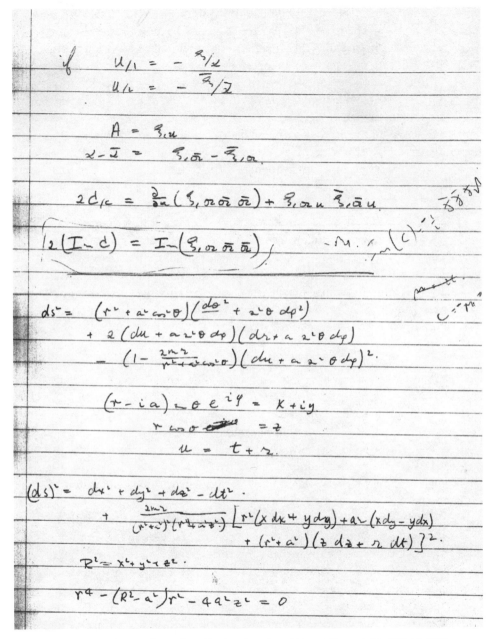

$$if \quad u_{/1} = -\frac{3}{z}/z$$
$$u_{/2} = -\frac{\bar{3}}{z}/\bar{z}$$

$$A = 3_{,u}$$
$$\varkappa - \bar{J} = 3_{,\bar{\sigma}} - \bar{3}_{,\sigma}$$

$$2 c_{/\kappa} = \frac{\partial}{\partial u}\left(3_{,\sigma\bar{\sigma}}\bar{\sigma}\right) + 3_{,\sigma u}\,\bar{3}_{,\bar{\sigma}u}$$

$$\frac{1}{2}\left(I - c\right) = I_m\left(3_{,\sigma}\bar{\sigma}\bar{\sigma}\bar{u}\right) ,$$

$$ds^2 = \left(r^2 + a^2\cos^2\theta\right)\left(\frac{d\theta^2}{} + \sin^2\theta\, d\varphi^2\right)$$
$$+ 2\left(du + a\sin^2\theta\, d\varphi\right)\left(dr + a\sin^2\theta\, d\varphi\right)$$
$$- \left(1 - \frac{2mr}{r^2 + a^2\cos^2\theta}\right)\left(du + a\sin^2\theta\, d\varphi\right)^2 .$$

$$\left(r - ia\right)\sin\theta\, e^{i\varphi} = x + iy$$
$$r\cos\theta = z$$
$$u = t + r .$$

$$\left(ds\right)^2 = dx^2 + dy^2 + dz^2 - dt^2 .$$
$$+ \frac{2mr}{\left(r^2 + a^2\right)\left(r^4 + a^2 z^2\right)}\left[r^2\left(x\,dx + y\,dy\right) + a^2\left(x\,dy - y\,dx\right)\right.$$
$$\left. + \left(r^2 + a^2\right)\left(z\,dz + r\,dt\right)\right]^2 .$$

$$R^2 = x^2 + y^2 + z^2 .$$

$$r^4 - \left(R^2 - a^2\right)r^2 - a^2 z^2 = 0$$

FIGURE 7.2. A second page from Kerr's notes where his spacetime solution (the equation for the interval *ds* near the middle) was first derived. This is the principal result that appeared later that year (1963) in his breakthrough paper. (Kerr's notebook containing this page is held in the Alexander Turnbull Archives of the National Library of New Zealand, Wellington.)

immediately that he was on to something quite special when he looked to see what would happen to his spacetime for an observer far from the object that creates it. In mathematical parlance, his solution was asymptotically flat. This means that the distant observer sees a smaller and smaller effect as he or she recedes from the central object—exactly as one would expect for a localized source of gravity whose influence dies off according to Newton's law far from the origin (fig. 7.2).

But was this really the spacetime surrounding a spinning object—possibly even something that John Wheeler would later call a *black hole?* The next day Kerr walked into Alfred Schild's office, the director of the new Center for Relativity in Austin, and described what he had done.

After morning coffee the two retired to Kerr's office, their steps more upbeat than usual. Back then Kerr occupied a modest little room on the ground floor, with a window facing south toward the state capital. Sometimes he imagined that he could actually see part of the dome through the overgrown bushes outside the building. He sat at a secondhand desk with his back to the door, opposite the window. To his right, he kept a worn-out armchair, where Alfred was now sitting. At this time of day, the sun filtered through the leaves and splashed slanted triangles of light across the paneled walls.

There were no computers back then; a black telephone provided the only electronic contact with the outside world—if one could find it. Kerr maintained what one would call a "busy" office: books piled up on top of papers, on top of books. Empty pizza boxes (and some not so empty), cigarette stubs, and Coca-Cola bottles filled up whatever space was left.

Both men smoked heavily, Kerr burning through several packs of cigarettes a day, while Schild found meditative comfort in his pipe. With both men now puffing away, the room took on a grayish tinge and a hint of gloom.

Alfred was a kind and cheerful man, with a distinguished flock of silvery hair (fig. 6.12). Though he was given to moods of depression when things were not going well, he always had a happy demeanor toward his colleagues and friends. He had come to Kerr's office on this occasion to be there with him as he calculated the angular momentum of the object producing the spacetime he had derived the previous day. Objects that do not spin have zero angular momentum. Any inferred nonzero angular momentum meant that the source of gravity was spinning—the "holy grail" of general relativity since the theory's inception many decades earlier. "It would take only twenty minutes," Kerr had promised over coffee that morning.

Like Schwarzschild's solution many years before, the Kerr spacetime has a singularity—not a point, as in the former case, but a ring (see fig. 8.6). It would become apparent later that this was due to the angular momentum sustaining his object from collapsing all the way to the middle. Instead, the size of the ring depends on how much angular momentum it possesses. An analogy is provided by a planet's motion in our solar system. Planets with large angular momentum orbit far from the Sun; those with smaller angular momentum orbit progressively closer to the center. Were our planet to somehow lose all of its angular momentum, it would fall directly into the middle, as would the rest of the material in the Sun's vicinity under such circumstances.

Directly measuring the angular momentum of a ring singularity is impossible. However, general relativists had known since 1918 that a rotating source of gravity would drag the spacetime around itself.[2] And the degree to which the spacetime spun was in turn a measure of how rapidly the central object was rotating. Kerr knew that by looking at his spacetime far from the center, he could determine whether or not *inertial* frames were being dragged and, even more importantly, at what rate they were forced to swirl.

While Schild waited patiently in the armchair, Kerr began calculating at his desk. Hunched over his notes, he was deep in thought; not a word was spoken for almost half an hour. But nothing needed to be said—until Kerr put down his pencil and looked up. "Alfred, it's spinning."

"Its angular momentum is Ma," he continued. The constant a was one of the four parameters that cropped up in Kerr's solution. Two of the other quantities were unimportant and ignorable. The fourth was M, the mass of the central object.

Schild jumped out of his chair, beaming. He appeared to be far more excited than Kerr himself and clearly knew what this meant. Cutting through the billowing smoke from his pipe, he rushed to the desk and looked over Kerr's shoulder at the scratchings on the table. By this stage it was a simple calculation, one that any relativist could have done once the metric was derived. For the first time since Einstein and Hilbert published their field equations of general relativity back in 1915, someone knew without a shred of doubt that these expressions possessed a solution describing spacetime surrounding a spinning object. Kerr stood up, and the two men walked toward the door. They headed out to lunch to celebrate one of the most remarkable achievements of the twentieth century.

---

2. This phenomenon is commonly referred to as *frame-dragging*, an effect first derived in 1918 from the theory of general relativity by the Austrian physicists Josef Lense (1890–1985) and Hans Thirring (1888–1976). It is also known as the *Lense-Thirring effect*. See Lense and Thirring (1918).

The one-and-a-half-page paper describing Kerr's result was submitted for publication on July 26, 1963, once he had returned to the Wright-Patterson Air Force Base in Dayton, Ohio. It was accepted right away and appeared in the journal a month later.[3] Its processing was so rapid, in fact, that several typographical errors remained in the final published version. Fortunately, this defect did not seem to matter. In the four decades since the paper's release, not a single person has ever contacted Kerr to mention they had found a mistake.

Two decades later the Nobel Laureate Subrahmanyan Chandrasekhar (1919–1995) (see figs. 7.3 and 10.2) would write about this discovery in his book on truth and beauty in science, in which he offered the following assessment:

> In my entire scientific life, extending over forty-five years, the most shattering experience has been the realization that an exact solution of Einstein's equations of general relativity, discovered by the New Zealand mathematician, Roy Kerr, provides the absolutely exact representation of untold numbers of massive black holes that populate the universe. This shuddering before the beautiful, this incredible fact that a discovery motivated by a search after the beautiful in mathematics should find its exact replica in Nature, persuades me to say that beauty is that to which the human mind responds at its deepest and most profound.[4]

The reaction from the general relativity community was immediate and very positive. This is what everyone had been waiting for. Einstein's code had been broken. The University of Texas immediately offered Kerr a tenured associate professor position, bypassing the usual untenured assistant professor route. But he first wanted to wait and see what would be happening in Dayton. As it turns out, Goldberg himself had been offered a professorial appointment with Bergmann's group in Syracuse, and so the two agreed to move on, one heading north to upstate New York, the other back down to Texas for a permanent position.

Riding a wave of excitement and a sense of desperation to complete the work, Kerr planned out a schedule in his mind that would keep him occupied for several years. But he had never been patient enough to write down everything he had done, and the brief paper announcing his discovery said very little of the beautiful mathematics he had developed to achieve his historical result. This inability to publish frequently and reliably would be a constant source of torment for the man who could always come up with

3. See Kerr (1963).   4. See Chandrasekhar (1987).

FIGURE 7.3. The astrophysicist Subrahmanyan Chandrasekhar made numerous contributions to the study of compact objects, including black holes. He would be one of the first users of the Kerr solution to study the spacetime surrounding a spinning object. (Image courtesy of the University of Chicago News Office.)

far more insight and results than he could reasonably put down on paper. But the impact of his work was simply too great for others to ignore, and it would be only a matter of months before he discovered firsthand how quickly his results could inspire others to push the investigation forward in other directions.

This happened upon Kerr's return to the UT campus in the fall of 1963, which coincided with the arrival of Roger Penrose (fig. 7.4) at the Center for Relativity. Schild's efforts at building up the group continued with the impressive recruitment of the world's best young relativists.

Discussions between Kerr and Penrose developed quickly, facilitated by the proximity of their respective offices. Kerr had continued to play with Einstein's equations, deriving interesting results that did not appear to have any immediate impact. But one of these was to become an important component of Penrose's lifelong work. From his solution of Einstein's equations, Kerr had found a simple way of characterizing all bundles of shear-free light rays (see fig. 5.4) in any given spacetime.

Penrose was impressed. The next morning he held an impromptu gathering in his office, with everyone in the Center for Relativity in attendance, listening to his description of this new result, which in time would become known as the *Kerr theorem*. Kerr himself never thought it was important

FIGURE 7.4. Roger Penrose at the International Astronomical Union meeting of 1973. (Image courtesy of Andrzej Trautman and the photographer Marek Holzman.)

enough to bother publishing it, but like Fermat with his last theorem,[5] it was commonly acknowledged that he had in fact proven it. The theorem was eventually published about a decade later by other investigators.[6]

Meanwhile, the momentum generated by the rapid development of the Center for Relativity in Austin galvanized the relativists' desire to foster closer ties to the astrophysics community, particularly now that Maarten Schmidt's discovery of quasars had demonstrated the likely importance of general relativity in explaining real phenomena in the cosmos.

In one of history's most inspiring coincidences, the same year that Kerr had finally cracked the Einstein code, Schmidt had been pondering over the nature of a starlike object with truly anomalous characteristics at the Palomar Observatory in California.

5. Fermat's last theorem is one of the most famous theorems in the history of mathematics. It simply states that it is impossible to separate any power higher than the second into two like powers, or that the simple equation $a^n + b^n = c^n$ has no solution for nonzero integers a, b, and c. Though simple to understand, this theorem could not be proven for 357 years, even if Fermat himself claimed to have found a proof that would not easily fit in the margin of his notebook.
6. Ironically, the Kerr theorem was published by other individuals at the Center for Relativity in Austin, after Kerr had returned to his native New Zealand. See Cox and Flaherty (1976).

The astronomical puzzle on Schmidt's desk was the star recently associated with the 273rd entry in the third Cambridge catalog of radio sources, known as 3C 273. Such sources had gone unnoticed by astronomers for centuries, appearing in the nighttime sky merely as faint points of light. But the growing power of radio telescopes in the 1940s eventually led to observations showing that several regions of the cosmos are very bright emitters of centimeter-wavelength radiation.

Just prior to 1963, the British astronomer Cyril Hazard had cleverly devised a method of pinpointing the exact location of such a source. Using the Moon's motion across a radio-bright region, he proposed, it should be possible to note the precise instant that the radio signal stopped and then reemerged as the Moon passed in front of it. A comparison with optical photographs of the sky would then show with which, if any, of the known astronomical objects in the firmament the radio emitter could be identified.

Hazard made arrangements to carry out his proposed observations with the Parkes Radio Telescope, situated in the Australian outback, several hundred miles from Sydney. The poor fellow took the wrong train out to the facility that day and missed the event entirely. But the telescope staff, headed by John Bolton (1922–1993), wouldn't let this opportunity pass and went ahead with the plan anyway.

However, there were even more serious problems to address. It turned out that the source was too close to the horizon, and the telescope could not tip over sufficiently to make the recording. Undaunted, the observatory staff cut down the intervening trees and removed several safety bolts from the structure, permitting it to swivel sufficiently to catch the occultation.

The experiment worked to perfection, and the radio source being tracked that night—3C 273—could be identified with a single starlike object in the constellation Virgo. Its docile appearance belied the fact that this quasi-stellar radio source (hence the origin of the name *quasar*) was a prodigious emitter of radiation.

What was particularly puzzling to Schmidt about 3C 273 was the spectrum of light it produced. The colors simply did not match anything ever seen before on Earth. Schmidt's enduring achievement that year was his eventual realization that the pattern of light before him was actually that produced by hydrogen atoms, only with wavelengths shifted by about 16 percent from their value measured in the laboratory.[7]

This shift was quite remarkable, for it had been known since the time of Edwin Hubble (1889–1953) that cosmological distances scale directly with

7. Maarten Schmidt reported his discovery in a one-page article published by *Nature* in 1963.

speed. An object's redshift is a direct measure of the rate at which it is receding from us. Thus, according to the wavelengths Schmidt had calculated, 3C 273 had to be much farther away than had previously been imagined, and must therefore be very energetic in order for us to see it at such a great distance.

Understandably, the astronomical and astrophysical communities were buzzing with speculation that year. Quasars were unquestionably the most powerful objects in the universe, and none of the known physical processes could power their potent output. What were they?

Ivor Robinson, Alfred Schild, and Englebert Schucking took this opportunity to organize a conference, held on December 16–18, 1963, in Dallas, Texas, that was to become the seminal gathering in a series of meetings that have grown in scope and size ever since. This First Texas Symposium on Relativistic Astrophysics brought together three hundred relativists, astronomers, and astrophysicists, all seeking to learn how a cross-disciplinary approach to the study of quasars could further our understanding of high-energy phenomena through the prism of relativity. The sentiment at that time was perhaps best expressed through the thoughts of Thomas Gold of Cornell University, who, in an after-dinner speech at the meeting, offered the following perspective:

> [The mystery of the quasars] allows one to suggest that the relativists with their sophisticated work are not only magnificent cultural ornaments, but might actually be useful to science! Everyone is pleased: the relativists who feel they are being appreciated and are experts in a field they hardly knew existed, the astrophysicists for having enlarged their domain, their empire, by the annexation of another subject—general relativity. It is all very pleasing, so let us all hope that it is right. What a shame it would be if we had to go and dismiss all the relativists again.[8]

Arriving at the Texas Symposium in Dallas, Kerr was informed that Penrose, a more accomplished speaker than him, had been asked to talk to the assembly about the Kerr solution. This was a rather strange decision, and, understandably, Kerr complained furiously to the organizers, who immediately rectified the situation. Though Penrose might have related better to the astrophysicists in the audience, Kerr would have found it difficult to sit there and listen to someone else present his groundbreaking work.

The symposium was packed full of talks, presentations, and informal discussions in the hallways and in the rooms. The quasar discovery had fueled much speculation, including the possibility that the odd, gargantuan

8. See Gold (1965).

objects known variously as *frozen stars* or *dark stars* (termed *black holes* later in the decade) had finally been discovered. So packed was the meeting, that talks began at 8:30 in the morning and continued throughout the day until 6:00 in the evening. Discussions and arguments continued long after that, well into the early hours of the morning.

Kerr's presentation was a ten-minute talk slipped in during one of the sessions. He was unknown then, except to his immediate colleagues and friends from the Center for Relativity in Texas. Of the three hundred participants, perhaps fifty or sixty were relativists; the rest were astronomers and astrophysicists.

Eventually, the Kerr solution would be recognized as the complete description of spinning black holes, which are now believed to be powering the most potent objects in the universe. But most of the audience had come to Texas to hear about quasars. They knew little, if anything, of general relativity, and certainly nothing about Kerr's esoteric mathematical arguments.

As Kerr got up to speak, many took this opportunity to stretch their legs or to continue their informal discussions outside the lecture hall. Many of those who stayed behind paid little or no attention. Only the relativists focused on what he was saying and were clearly annoyed at the lack of appreciation of Kerr's work by the majority of those in attendance at this meeting.

Achilles Papapetrou (1907–1997), one of the world's leading relativists, could not bottle up his frustration any longer. At the end of Kerr's presentation, he took the floor and gave an impassioned admonition to the audience. He explained how he, Papapetrou, and most of his colleagues had worked for thirty years trying to find a solution to Einstein's equations, failing repeatedly. They knew that unless a solution could be found describing systems with realistic properties, it would be difficult to perceive general relativity as being fully relevant to the physical world.

The audience gave Papapetrou's remarks a polite reception. The next speaker redirected the discussion back to quasars, and the audience perked up again, the topic now having reverted back to what they had come to hear at the meeting.

8 :: BLACK HOLE

Back in the 1960s—in the midst of the golden age of relativity—the eager young stars who considered themselves members of this growing community were among the most brilliant scientists of their day. And it took the intellectual power of many of them to fully comprehend the implications of the Kerr solution. But they worked fast, and by the time John Wheeler coined the term *black hole* in 1967, many of the properties of these exotic objects had been identified and analyzed. Unfortunately, the end of this period (chapter 9) would also be tainted with an entirely unpredictable tragedy.

Kerr himself realized immediately that his solution had the same kind of surface (that we now call an *event horizon*) recognized long before in the Schwarzschild metric. When we write down the latter in coordinates appropriate to an observer at large distance from the central object—that is, when we cast the Schwarzschild spacetime in terms of rulers and clocks we can use at our remote location—we encounter a limitation on how close to the center we can see phenomena developing over a finite time. As discussed earlier in this book (chapter 4), the great acceleration close to the source of gravity slows down time, which according to our clocks appears to stop completely as we reach the so-called Schwarzschild radius.[1]

Of course, another observer—perhaps one who finds himself falling haplessly inward—makes measurements using his own rulers and clocks, which are not affected by acceleration (according to him), so there is no limitation to how close to the center he (or his remains) can get. In preparation for the First Texas Symposium in Dallas just after his solution was pub-

---

1. Formally, this occurs at $2GM/c^2$, where M is the mass of the black hole, G is Newton's gravitational constant (expressing the strength of gravity), and c is the speed of light. If the Sun were to collapse into a black hole, its Schwarzschild radius would be roughly three kilometers. For a 10 million solar-mass object powering a typical quasar, this radius increases to about 30 million kilometers, roughly one-fifth the distance between Earth and the Sun.

lished, Kerr examined the trajectories such an observer could follow and discovered, not surprisingly, that after you crossed a certain radius, there was no possible return.

As was his custom, however, once Kerr solved a problem in essence, he was ready to move on impatiently to the next challenge. But there was much more to this particular problem than he had realized, beyond simply knowing that there was an event horizon. Indeed, although Kerr correctly obtained the "size" of his black hole as seen along the spin axis—that is, in the direction about which the central object spins—the formula he published in the conference report for his horizon's radius was incorrect for all other directions.[2]

As we now know, not only is the horizon for a spinning black hole not the same as that for Schwarzschild, but it is not even a sphere when one uses the same coordinates as for the former. There are in fact several different horizons, and their impact on a nearby observer depends on whether or not he is moving relative to the center. An understanding of this phenomenon—and the structure of black holes in general—would unfold over the next four years from the combined effort of several people.

The week after the meeting in Dallas, Alfred Schild hosted a Christmas party at his house in Austin, but neither he nor Kerr could refrain from injecting relativity into their conversation. Inevitably, the chatting turned into a serious discussion, and the two retired to Schild's study, abandoning the holiday revelry behind them.

The solution Kerr had found earlier that year was but one form of a general group of solutions all satisfying the symmetries and simplifications he had cleverly guessed in order to reduce Einstein's equations to a manageable form. That evening, on a blackboard in Schild's house, the two men calculated all the rest, now known generically as the Kerr-Schild metrics.[3] The Kerr solution is itself a member of this set, though it is usually referenced separately because of its historic importance and uniqueness.

This family of solutions is interesting but, for the most part, not very relevant to nature. The reason is that, whereas the Kerr metric itself describes the spacetime surrounding a localized object—with a central ring singularity (see fig. 8.6) encased by an event horizon—the rest of the Kerr-Schild

2. The expression Kerr wrote for his horizon's radius reduces to the correct value along the black hole's spin axis but shows an incorrect dependence on the coordinates, which invalidates it for any other direction. See Kerr (1965).

3. One cannot help but notice the amusing similarity in names between the Schwarzschild and Kerr-Schild metrics. Even more, *Schwarz* means "black" in German and *schild* is "shield," a fitting label ("black shield") for a solution describing the spacetime around a black hole.

metrics all contain singularities that extend to infinity. We know of nothing in the universe that has such an abstruse geometry. One cannot even say if they pertain to a spinning object (or objects), since one cannot calculate their angular momentum.

Nonetheless, the Kerr-Schild collaboration provided one additional benefit to the scientific community, for it ensured that the extensive mathematical work leading up to these results would find its way into the published literature. Left to his own, Kerr would not have bothered writing up the steps he followed. But Schild was much more disciplined in this regard, and in 1965 the two published a long, detailed description[4] of all the beautiful calculations carried out since late 1962. In 2007 their paper on this topic would be designated a "Golden Oldie" by the journal *General Relativity and Gravitation*.

The meeting in Dallas had done little to help the astrophysicists embrace the relativists and their work, but even this could not quell the frenzy with which those who understood the significance of Kerr's work began to study its meaning and consequences. Chief among them was Ted Newman (fig. 8.1), who earlier that year had seemingly abandoned the chase for a meaningful solution to Einstein's field equations describing a realistic object.

At the beginning of 1964, Newman set himself the task of finding an even more general solution than the one uncovered by Kerr the previous year. The Kerr metric, like Schwarzschild's, describes spacetime surrounding a highly concentrated object. A key requirement in their solutions is that the region surrounding the source of gravity is a vacuum. Without this simplification, Einstein's equations are virtually unsolvable, except in certain computer simulations. The field equations describe how the gravitational influence changes when one passes through a region where matter and/or energy may be present. By adopting vacuum conditions, Schwarzschild and Kerr simply ignored any contribution to gravity from everything outside the central object.[5] They could therefore set one side of Einstein's equations equal to zero, which forced the spacetime to vary only in response to the mass at the center.

It was realized soon after the Schwarzschild solution was published that, although finding a description of spacetime around an object surrounded by matter was incredibly difficult, one could nonetheless do away with the vacuum condition and still get something useful. Instead of vacuum, one could fill the space with the field of force created by an electric charge.

---

4. This work was reported in Kerr and Schild (1965).
5. This approach is similar to the way we calculate the gravitational acceleration on a projectile above Earth's surface. To make this determination, we include only Earth's mass, ignoring its atmosphere, whose mass is negligibly small compared to that of the planet itself.

FIGURE 8.1. Ted Newman (*on the left*) talking with Peter Bergmann at the International Astronomical Union meeting of 1973. (Image courtesy of Andrzej Trautman and the photographer Marek Holzman.)

All of the matter with which we are familiar is composed of charged particles. An atom, for example, contains a positively charged nucleus surrounded by a negatively charged electron cloud. If the temperature of a hydrogen gas is raised to 10,000 degrees or more, the nuclei and electrons receive sufficient energy to separate from each other, creating a plasma. It is even possible under certain circumstances to separate the different charges spatially, creating a strong field of force between them.

But this field of force carries energy, and according to Einstein's theory of relativity, energy is a source of gravity, just like mass is. Thus, if a Schwarzschild black hole were to absorb matter dominated by one charge (say the nuclei of dissolved hydrogen atoms), then a field of force would emerge between the collapsed object and the oppositely charged particles (in this case, electrons) on the outside. Such a field is relatively easy to describe as a source of gravity, and soon after Schwarzschild published his groundbreaking work, a similarly simple solution was found for the spacetime surrounding a *charged* black hole.

It should be possible to do the same thing for a charged, spinning object, Newman thought, and he set about finding the more general solution. His idea was rather straightforward. Now that the Kerr metric was known, he could start with that form of the solution and guess how the spacetime might be modified in the presence of a charge. He wrote out a series of trial expressions and then assigned them as an exercise to students[6] in his course on general relativity. Their homework was to see if any of these guessed solutions actually satisfied Einstein's field equations. Amazingly, one of them did!

What Newman did not know at the time (but perhaps speculated) was that Kerr and Schild were themselves working on a generalization of (not only the Kerr metric itself, but) the whole class of Kerr-Schild metrics. In 1964 Kerr had assigned part of this problem to his own student George Debney, and the three of them would eventually publish the work years after they themselves had completed it, their sense of urgency having waned after they heard about Newman's result.[7]

Today we believe that the solution based on the Kerr metric, with the possible inclusion of charge, represents the most likely configuration of astrophysical black holes, about which we will have more to say in chapter 11. Known as the Kerr-Newman metric—in honor of the breakthrough first made by Kerr, then followed by Newman's generalization to a charged configuration—this description of the asymptotically flat spacetime surrounding a spinning object requires the specification of just three numbers: the object's mass, its spin, and its charge. This result, which Wheeler humorously described as "a black hole having no hair," would be understood after the painstaking and insightful work carried out over the subsequent four years.

It is indeed remarkable how quickly the young relativists pounced on the black hole problem. Far from trivial, the question concerning how many different types of black holes nature permits requires a deep understanding of Einstein's equations. Barely four years after Kerr published his groundbreaking solution, the Canadian Werner Israel (see fig. 8.2) announced the first so-called black hole uniqueness theorem, proving that the only type of nonspinning black hole possible was that found by Schwarzschild sev-

---

6. These were E. Couch, K. Chinnapared, A. Exton, A. Prakash, and R. Torrence.
7. A preliminary announcement of this result was published in Kerr and Schild (1965). The full description appeared in Debney, Kerr, and Schild (1969).

FIGURE 8.2. Born in Berlin in 1931, Werner Israel (shown here circa 1967) was raised in South Africa but eventually settled near the University of Alberta, Canada. (Image courtesy of Werner Israel.)

eral decades earlier.[8] In subsequent years work by several others, including Brandon Carter (fig. 8.7) and Stephen Hawking, allowed Israel's pioneering result to be generalized to the case of a spinning object, proving that the Kerr metric (or Kerr-Newman metric if a global charge is present) is the unique description of all spinning black holes.

These striking results demonstrate that no matter what shape a star may have as it collapses under its own weight, by the time it attains an event horizon and becomes a black hole, it must have (at most) only three distinguishing features—hence the no-hair analogy coined by Wheeler. Any such object may be identified solely on the basis of its mass, its electric charge, or its spin. No other specific features may be present. So how does a black hole become so simple? Chapter 4 introduced the idea that a variable source of gravity may emit gravitational waves. A collapsing object with a

8. Israel's argument is elegant and relies simply on the assumption that the black hole has reached its equilibrium configuration, with a high degree of spatial symmetry. The technical details may be found in his paper "Event Horizons in Static Vacuum Space-Times" (1967).

complicated geometric structure will radiate with greater intensity as it approaches its terminal state, in the process erasing any peculiar features it may have originally possessed. Viewed from our distant vantage point, only its mass, charge, and spin remain as observable qualities.

In nature, however, it is extremely difficult to separate charge over great distances. The electric (or more generally, the electromagnetic) force is many orders of magnitude greater than gravity, so any movement of charge is associated with strong nongravitational influences.[9] Though the universe can in principle contain Kerr-Newman black holes, it is therefore far more likely that when we do identify a black hole somewhere in the cosmos, it will "simply" be a Kerr.

The development of the Kerr-Newman spacetime came at the beginning of a period in which the properties of event horizons were being studied seriously, recognizing that they might actually be relevant to real objects in nature. There are probably several reasons why this transition was delayed until the 1960s—almost half a century after the foundation of general relativity. No doubt, Einstein's reluctance to accept the reality of such oddities had a lot to do with this, and after his death in 1955, the next generation of relativists felt fewer restraints on their exploration.

It could also be said that prior to this period, most of the individuals who became interested in relativity lacked the deep mathematical insight and background brought to the table by the generation that followed Einstein, best exemplified by Kerr and Penrose themselves, who began their careers as mathematicians and evolved into mathematical physicists. Even Einstein was somewhat out of his league when it became necessary to formalize his theories, and he relied heavily on the mathematical expertise of his friends and colleagues, including Marcel Grossmann (1878–1936) and David Hilbert (see chapter 3), to help him produce the final version of his gravitational field equations.[10]

Indeed, it would be Penrose's mathematical virtuosity, soon after Kerr discovered his historic solution, that would help him produce one of the most important black hole theorems. Like Kerr, Penrose was drawn at a young age by the elegance and beauty of mathematics. But on one of his trips to Cambridge, he met Dennis Sciama (1926–1999), who would later supervise the graduate research of many luminaries in general relativity and

9. As a concrete example of this disparity between the electromagnetic and gravitational forces, consider the fact that even a small bar magnet is sufficient to (electromagnetically) lift a piece of metal from the table, against the restraining (gravitational) force applied to it from the whole Earth beneath it.

10. Grossmann and Einstein were classmates. When Einstein sought to formulate mathematically his ideas on the general theory of relativity, he turned to Grossmann for assistance and from him learned the basic elements of differential geometry.

cosmology, including Brandon Carter, Stephen Hawking, and the current Astronomer Royal of Britain, Sir Martin Rees. Though Penrose would never work for Sciama directly, the two men discussed astrophysics often, and Penrose became inspired to ply his mathematical training in the field of general relativity.

In the fall of 1964, shortly after Penrose's departure from the Center for Relativity in Austin, where he had learned firsthand about Kerr's revolutionary work, he was walking back to his office at Birkbeck College, London, wondering whether singularities were an unavoidable consequence of the Schwarzschild and Kerr spacetimes. His thoughts drifted away from the usual equation-driven description of such structures and settled instead on a more geometrical, shape-driven conceptualization, drawn from another important branch of mathematics known as *topology*.

Rather than concerning itself with how quantities change from point to point, which is what Einstein's field equations do, topology is instead a discipline that attempts to uncover the connections between those quantities. For example, suppose we were to take a long strip of paper and lay it flat on the table. A topologist might wonder how many different surfaces that strip possesses. Assuming it is infinitely thin, the answer would be two—one on top, the other on the bottom.

However, let us next imagine twisting the strip once and gluing its ends together to form a *Möbius strip* (see fig. 8.3). Its topology would now be entirely different because it would contain just a single surface. An ant starting to move down the strip would be able to crawl along it indefinitely without ever leaving its surface. In this sense, the manner in which any pair of points on the strip of paper is connected would be very different in the two disparate cases.

The argument Penrose made was based on a few straightforward conditions, but in essence, his point was that if an event horizon forms, then at least one possible geodesic is incomplete, presumably because it must terminate at some point—a singularity. In other words, at least one trajectory (like a path along the Möbius strip) would always be bounded by the horizon due to the insurmountable curvature produced by the strong gravity. It would be as if one hurled a projectile higher and higher above Earth's surface, trying to free it from the planet's clutches. But with only limited strength, the hurler could never force the projectile to break free.

Gravity is always attractive, never repulsive.[11] So all the geodesics trapped within the event horizon eventually converge toward each other;

---

11. This was an important assumption in Penrose's theorem. Today we realize that this may not be true for some forms of matter and is in fact false for dark energy, the agent providing an accelerated expansion of the universe.

FIGURE 8.3. A Möbius strip is a band that has only one surface and a single edge. An ant crawling along it could continue its journey indefinitely without ever reaching an end.

and this must happen over a finite amount of time on the observer's own clock. The point of convergence is presumably the singularity inside the event horizon.

But we have to be careful how we interpret this theorem, because it is all based on the so-called *proper* time. Though it is true that a singularity forms within a finite amount of proper time since the development of an event horizon, the passage of time in the universe outside is actually infinite, due to the gravitational redshift effects we discussed earlier. Regardless of its interpretation, however, Penrose's singularity theorem was one of the first major advances to occur during the period (1963–68) in which the structure of black holes would be dissected and analyzed in great detail.

Kerr's breakthrough solution eased the anxiety many felt in using general relativity to describe what happens when stars collapse at the end of their lives. Stars spin, and their angular momentum does not go away when this catastrophic implosion takes place. Knowing that Einstein's equations allowed a description of spacetime around such an object gave people confidence that the implementation of general relativity in their calculations and analysis would lead to results of direct relevance to the real universe.

Mathematical physicists, such as Penrose in England, concerned themselves primarily with the structure of the spacetime inside and outside the event horizon. But other physicists, more interested in the direct astrophysical application of the Kerr solution, also took notice. Across the Atlantic, John Wheeler (fig. 8.4) and his group at Princeton had been very interested in the physics of neutron stars, already believed at that time to be one of the possible end points of stellar evolution. These are city-sized objects, though with a mass comparable to that of the Sun, supported from

FIGURE 8.4. John Wheeler (shown here circa 1970) helped to formulate the theory of gravitational collapse in the 1960s. In December 1967 he coined the term *black hole* in his public lecture "Our Universe: The Known and Unknown." This name caught on much more easily than the previously used name *frozen star*. (Photo by Blackstone-Shelburne, courtesy of Princeton University Archives, Department of Rare Books and Special Collections, Princeton University Library.)

further collapse by a quantum-mechanical effect.[12] These stars are mostly comprised of neutrons, which by virtue of how much internal angular momentum they possess must be oriented in such a way that no two of them can occupy the same space if their energies are identical.

Originally called the *Pauli exclusion principle*, this phenomenon can be understood physically as arising from a cancellation of their wave functions in quantum mechanics, due to the fact that a rotation by 360 degrees of a pair of these particles actually does not bring them back to exactly the same state they were in to begin with. No one yet understands why this happens, but it is an observed property of the three-dimensional space we live in.

12. There are many worthwhile references one may access for a more complete description of this phenomenon. A nontechnical account may be found in Melia (2003).

As a result of this "exclusion," the neutrons inside such a star cannot all collapse to the middle; they effectively repel each other and, in so doing, support a neutron star against total collapse to something else. Aside from clearly exhibiting this quantum-mechanical phenomenon, neutron stars are also fascinating to someone interested in general relativity because the support they receive endows them with a size barely larger than the event horizon they would have if they collapsed all the way to a black hole. The typical neutron star radius is about 10 kilometers; if all of its mass were confined to a singularity, it would be surrounded by an event horizon with a Schwarzschild radius of 3 kilometers—smaller, but not by much.

Wheeler and his group were therefore keenly following what was developing in general relativity at that time, particularly the exciting work involving event horizons. He decided to change the focus of their research from the physics of neutron stars to a study of the total gravitational collapse of objects too massive to resist a complete implosion toward a singularity.

This problem actually originated back in the 1930s, when J. Robert Oppenheimer and his collaborators had begun to investigate in earnest the evolution of stars beyond the point where much of their nuclear fuel is spent. While on the main sequence, meaning the period during which the star burns hydrogen-rich fuel at a steady rate, the ashes collect within its core, which forms a growing reservoir of heavy elements such as iron, carbon, and oxygen. The star can retain this balance for billions of years because the energy released from nuclear burning can sustain buoyancy in the hot gas against the inward pull of gravity.

But the fuel eventually runs out. Oppenheimer, best remembered in history for his contribution to the Manhattan Project that developed the atomic bomb, is also admired scientifically for having been the first to show that the end point of this stellar evolution can result in the catastrophic collapse to something approaching a singularity. He was challenging physicists everywhere to accept the absurd notion that dead stars collapse without limit to an indefinitely small size and infinitely large density.

Scientists working in the East and the West had different names for these objects: the Russians called them *frozen stars*, while Western physicists were calling them *collapsed stars*. (This designation made some sense since the gravitational redshift becomes infinite at its horizon, "freezing" the observed time there.) Neither designation fully described these objects, however, and it came to Wheeler to coin the term *black hole*, which was quickly and enthusiastically adopted by the astrophysical community.[13] Only the

---

13. Wheeler first used this term in December 1967, in a public lecture entitled "Our Universe: The Known and Unknown." The apparent menace implied by this name added enough intrigue to make it stand out from other terms that were being used at that time.

French and some Russians found exception to the term *trou noir*, which nonetheless found mainstream acceptance in due course.

Just prior to that year (1967), the term *black star* had also been used occasionally. In fact, this designation appeared in an early episode of the television series *Star Trek* (1966–69), a true cultural phenomenon in the United States and elsewhere around the world. But it is now a black hole we all think of when hearing the term *event horizon*.

The more relativists explored the Kerr black hole, the more they realized how complicated it is compared to Schwarzschild's. A particle approaching the latter experiences the same curvature from every direction. Since only gravity determines the particle's geodesic in this case, and gravity is isotropic, the event horizon looks the same from all directions, and there is only one such "surface."

As Lense and Thirring demonstrated in 1918, however, rotating sources of gravity drag the spacetime around them. And this swirling effect helps to support particles from falling directly inward under the influence of gravity, so by itself, this effect would move the horizon inward at the equator. But inside the black hole, the mass settles onto an equatorial ring, not a point singularity as it does in the Schwarzschild case. Thus, the location of the event horizon in the Kerr metric depends not only on the mass of the central object, but on its spin as well, since the force of attraction is now coming from the ring, not the center. Moreover, the curvature created by this ring is greatest at the equator, decreasing steadily toward the poles, so the horizon's shape is no longer spherical. Instead, it is a somewhat flattened sphere, with the widest portion in the equatorial plane (see figs. 8.5 and 8.6).[14]

Worse, the fact that there are now two influences determining the location of the Kerr event horizon means that there are actually *two* horizons (see fig. 8.6). This happens because the frame-dragging effect drops off faster with distance from the center than does gravity, so the light-speed condition is met twice. As the black hole's spin increases, the inner event horizon moves outward, and the outer one moves inward. Eventually, the two merge and vanish, leaving behind a "naked" (meaning that it is not encased within an event horizon) singularity that, as we have already pointed out, is a ring rather than a Schwarzschild point when the black hole's angular momentum is nonzero.[15]

14. The shape of the horizon depends on the type of coordinates used. The usual depiction of the Kerr horizon is as a sphere, but the radial coordinate in this representation is not the same as that used in the Schwarzschild metric. One sees the difference directly when the same coordinates are used.

15. In nature this could never happen, however, because a black hole being spun up by matter carrying angular momentum inward would ultimately drag the spacetime so

Schwarzschild Black Hole

Axis of Rotation

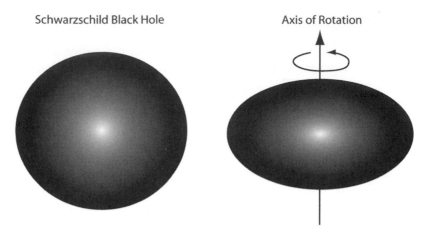

FIGURE 8.5. For a Schwarzschild (nonspinning) black hole, the event horizon (shown on the left) is strictly spherical. When the black hole is spinning, however, the singularity in the interior is a ring, not a point, and therefore the curvature it induces depends on angle, so the (Kerr "outer") event horizon (shown on the right) is widest at the equator, tapering off gradually toward the poles.

FIGURE 8.6. The Kerr black hole has two horizons: an "outer" horizon and an "inner" horizon, which approach each other as the object's spin increases, until they merge and vanish. The singularity at the center is a ring, rather than the point one sees in the Schwarzschild case. The size of this ring depends on how much angular momentum the black hole possesses.

By now it must be apparent how often we have made reference to particle trajectories in elucidating these black hole properties. This should not be surprising, of course, because in describing a black hole's influence, we care most about how it affects its environment. So while Penrose applied his mathematical skills to an evaluation of the black hole's interior, and Wheeler and his group simulated a stellar collapse into a black hole, other relativists—notably Brandon Carter—carefully calculated the particle trajectories (or geodesics) both outside and inside such an object. The Australian Carter had gone to Cambridge University to work with Dennis Sciama and was about to begin his doctoral research just as Kerr produced the now-famous solution. By the mid-1960s he was examining the structure of the Kerr metric in earnest.

Carter was one of the first to write down the Kerr spacetime in terms of the rulers and clocks used by an observer falling inward, rather than someone fixed at a large distance from the black hole. This allowed him to examine the particle trajectories through the two event horizons, all the way to the ring singularity in the middle. He discovered, in this fashion, one of the most controversial properties of Kerr black holes that begs explanation even to this day.

He found that, in addition to having a ring-shaped (fig. 8.6) singularity at its center (which Kerr had already noted in his 1963 discovery paper), the Kerr spacetime also possesses a small region near this ring that contains closed "timelike geodesics." What this means simply is that there is a region near the middle of the Kerr black hole within which an observer may follow a loop and come back to his starting point, either in the future or in the past, depending on the direction of travel.

Several writers since then, including Carter himself, have dubbed this a "time machine," since in principle an observer could move backward and forward through time by simply crawling in one direction or the other along these closed paths. It is not yet clear what to make of this, but as was the case with Penrose's singularity theorem, it is critical for us to place this result in its proper context. We must remember that while all this is happening within the black hole, an infinite amount of time is being expended by our universe on the outside. Moreover, these closed timelike geodesics are encased entirely within the inner Kerr horizon. As far as we know, we therefore have no way of testing any of these ideas and results from the outside.

It should also be pointed out that the closed timelike loops exist only if

---

rapidly around itself that the last vestiges of plasma required to push the black hole's spin above the threshold would simply be flung outward rather than be drawn in across the event horizon.

FIGURE 8.7. Brandon Carter, shown here at Kerr's celebratory meeting in 2004, was the first relativist to closely examine particle trajectories in a Kerr spacetime. (Image courtesy of Brandon Carter.)

the interior is comprised of the singularity surrounded by a vacuum. Any matter or energy, other than the ring, will alter the geodesics and disrupt the closed loops. So this situation is somewhat of a catch-22, in the sense that closed timelike loops may exist near the singular ring, but as soon as an observer attempts to use them, they vanish, since his or her presence affects the curvature and ruins the very phenomenon they came to experience.

However, this highlights another interesting property of Kerr black holes that distinguishes them in a very important way from Schwarzschild's. As we learned in chapters 2 and 3, the second postulate of special relativity—that the speed of light is constant in any given frame and also from observer to observer—means that the interval calculated between any two spacetime points[16] is an invariant. Everyone must agree on the length of that interval.

The adjustment of time and space needed to keep this interval invariant becomes extreme when a distant observer tries to examine what is happen-

---

16. This is essentially the length traversed by a ray of light in that frame over a given time interval, as measured by a stationary observer. If the observer is moving, then the interval is still the same, though the passage of time adjusts based on the distance he covers during that measurement.

ing close to the event horizon. In order to keep the interval unchanged, time (as measured by that distant observer on her clock) must become spacelike, and space (as measured with her ruler) must become timelike across that surface. This phenomenon, however, is observer dependent. For someone falling in freely, time and space function as uninterrupted continua in that frame, though the distant observer would infer that an infinite amount of time is required for the process to be completed. Perhaps the assumption of an invariant interval is overly restrictive, and this may be where general relativity needs to be modified. If so, the breakdown would occur not because of the theory itself, but rather because the second postulate of *special* relativity may not be valid in the presence of strong gravity, which gives rise to enormous accelerations.

In the case of the Kerr black hole, this feature of the interval has far-reaching consequences because, whereas in a Schwarzschild black hole, time and space adopt each other's characteristics once, in the case of a Kerr black hole, they do it twice. Thus, for a Schwarzschild black hole, one cannot avoid hitting the singularity in the middle once the event horizon is crossed. Every geodesic is terminal, meaning that within a finite passage of proper time (on the observer's clock), he or she will find themselves reaching oblivion in the middle.

But for a Kerr black hole, time and space adopt each other's characteristics upon crossing the outer horizon (see fig. 8.6), and then revert back to their original function again across the inner horizon. What that means is that if an observer were to survive the passage across these two event horizons, he or she would find themselves in a spacetime like that on the outside of the black hole except, of course, for the presence of the singularity in the middle.[17]

In this case, however, it is not necessary for all geodesics to terminate on that ring. Indeed, only the geodesics already in the black hole's equatorial plane are destined to hit the singularity directly. All other paths head inward but may continue in all sorts of geometric configurations, depending

17. In reading through this description of the black hole interior, the reader should be aware of the fact that what we are considering here is based entirely on the Kerr metric. However, a real black hole interior may not be this simple. For example, if the interior contains any mass or energy (besides the singularity), the geodesics there would not have the form described above, and the structure of spacetime might itself be altogether different. One of the earliest relativists to think about the uniqueness of spacetime metrics, Werner Israel (figure 8.2), was also one of the first to be concerned with black hole interiors, and he became an influential contributor to this field well past the golden age of relativity, writing several landmark papers over the subsequent forty years. A technical account of his work on the structure of black holes may be found in Israel and Poisson (1990).

on the initial conditions. This is why, incidentally, closed timelike loops are permitted inside a Kerr black hole, because not all of the geodesics are forced to terminate at the singular ring. So as bizarre as these objects appear to be, there does exist this intriguing possibility that if an object were to survive its infall across the two event horizons, it could continue to retain its integrity inside the black hole, functioning in a space and a time similar to those in the exterior.

Some three to four years after Kerr published his solution, there were enough relativists seeking to understand its consequences, that each benefited greatly from the others' progress. Carter's work with particle geodesics around black holes would soon inspire Penrose to uncover yet another important property of the Kerr spacetime. His geodesics indicated that the Kerr black hole not only has two event horizons, but must also contain another critical radius—now known as the *static limit*—farther out.

Below this radius, no propulsion system, no matter how powerful, would be capable of sustaining a particle's position—hence the designation *static limit*. However, the particle is not completely doomed yet, for the black hole's frame-dragging effect forces it into a co-rotation that helps partially resist the inward pull of gravity. The particle is forever swallowed up only when it crosses the outer event horizon. Once inside this region, its path is inevitably traced to the interior region near the ring singularity, and not even its orbital motion can prevent it from the inexorable pull inward across the inner event horizon. (But as we've noted, unlike the situation inside a Schwarzschild black hole, not all the geodesics in the Kerr spacetime must necessarily terminate on the singularity itself.) The outer horizon is therefore also known as the *stationary limit*.

Penrose was the first to recognize that this region between the static and stationary limits, later called the *ergosphere* (from the Greek *ergon*, meaning "work") had some special properties, by virtue of the forced co-rotation of any object or particle that entered it.[18] If a particle were to fall into the ergosphere and split, say by decaying into two other particles, under certain conditions one of the debris particles would fall farther inward, while the other would gain a sufficient amount of energy from the black hole to facilitate its escape (hence the use of the word *ergon*). This mechanism, known as the *Penrose process*, is therefore a method of extracting energy from a black hole, through an effective emission of particles.

The golden age of relativity was now arguably at its zenith, with fresh young faces continuing to enter the field. By the time Wheeler had coined

---

18. This name was coined by John Wheeler at Princeton, together with Remo Ruffini, one of the young colleagues in his group at the time.

the term *black hole*, Sciama's next student, Stephen Hawking, had just finished his degree at Cambridge. He too would be influenced early on by Kerr's work, during his visit to the University of Texas in 1966–67.

Hawking had come to talk about a new theorem he thought he had proven. When matter collapses, he reasoned, only two options were available to it: either a black hole forms, or it creates a closed timelike loop, like the one discovered earlier by Carter in the interior of the Kerr black hole. The proof of his theorem—that a black hole must inevitably form—rested on the argument that these closed timelike loops (i.e., time machines) are unphysical, leaving only one option—the creation of an event horizon.

But though Hawking had described this as a proof while at the Center for Relativity, in retrospect it must have been a conjecture—a reasonable guess that might be proved later. Kerr went away and tried to prove it himself, but as hard as he tried, he could not.

Kerr found something like it, but he could not prove that if a singularity does not form, an external time machine is an inevitable consequence. Instead, he proved that if there were no singularity, one would see a geodesic that got arbitrarily close to the starting point, but not exactly at that location—no time machine, but perhaps something close to it.

At the end of Hawking's stay, Kerr met him again at a social function and described what he had found. Hawking later modified his theorem and published it. A few years later, Kerr was correctly credited[19] for providing the concept of almost closed timelike curves in spacetime. Hawking went on to become one of the most influential black hole theorists, well past the golden age of relativity when the essential solutions of Einstein's field equations had been found. He and Penrose have since developed several important theorems pertaining to black holes, including those emerging in the Kerr and Schwarzschild metrics.

The 1960s constituted an exhilarating period in science, particularly in relativity, drawing the most brilliant young investigators into its fold. Competition among them was fierce, interlaced with bouts of frenzy, euphoria, and occasional disappointment. But through it all, each of them benefited from the others' work. And history would mark this era as the time when Einstein's code was finally broken.

19. See Hawking and Ellis (1973).

9 :: THE TOWER

In his letter of reconciliation to his friend and colleague David Hilbert (see chapter 3), Einstein revealed a detachment from "this shabby world" due, perhaps, to his sense of elation at uncloaking nature's secrets. Another contributor may have simply been the solitude that often visits scientists deep in thought. But though the most creative among us may appear to be absolved of life's prosaic offerings, even its pitfalls and occasional tragedies, they are human after all, and their work cannot help but be influenced by strange and unexpected events.

A farm boy from New Zealand, Roy Kerr was not prepared for the emotionally draining experience he and others at the University of Texas would endure amidst this period of great scientific discovery. The relativity community worldwide would sustain a major blow, forever tinctured by the consequences of an unfortunate incident. The sad loss of Karl Schwarzschild— cut down in his prime after his fundamental, but singular, contribution to general relativity—would be echoed decades later in ways we may never fully comprehend.

Ray Sachs had arrived in Texas along with Roger Penrose in late 1963. By the summer of 1965, he and Kerr had decided it was time to put their collective talents together and finally work out the interior Kerr solution. Let us be clear about what this means at the outset. Penrose, Brandon Carter, and others were at this time working out the properties of the Kerr black hole in which everything had reached its final (equilibrium) state. That means a situation in which the black hole's horizons are fully formed, and everything inside has settled down to the ring singularity.

Kerr always felt that his solution should be more than just a description of fully formed black holes. After all, as we learned in the previous chapter, a neutron star is an object hovering just barely above a black hole configuration—so close to it, in fact, that general relativistic effects must be evident near its surface. He wondered whether the Kerr metric might describe the spacetime surrounding such a star, even if its interior was not encased within an event horizon and, especially, if it did not contain the ring singularity.

The task he and Sachs set themselves was to find a solution to Einstein's equations describing spacetime, not in a vacuum, but within the matter distributed inside the star—a hopeless goal, as it turned out, since even today, four decades later, no such solution has been found.

But in order to even attempt this work, Kerr and Sachs had to find a way of recasting the exterior Kerr solution into a form using the right set of rulers and clocks that would permit them to *continue* the solution through to the interior. They actually knew of a way to do this, since Achilles Papapetrou, the fellow who spoke in Kerr's behalf at the First Texas Symposium back in 1963 (see chapter 7), had just written a paper showing that one could simplify the metric by "combining" a spatial coordinate with time, thereby creating a new coordinate. (Again, what this means in simple terms is finding a different observer, with his or her own clock and ruler, whose coordinates make spacetime look simpler than that of the first observer.)

They quickly cast the Kerr solution into its new form and took a break for a cigarette. But after returning to the blackboard, they realized in only minutes that what they were trying to do was nonsense. The properties of the Kerr metric outside the star said nothing about what the solution would be inside, since that depended entirely on how the matter is distributed there. Dejected, they abandoned the effort, not realizing at the time how important this new form of the Kerr metric would turn out to be.

That fall Robert Hamilton Boyer (1933–1966), another bright mathematical physicist, joined the Center for Relativity in Texas. His expertise in geometry had also drawn him to the problem Kerr and Sachs had worked on during the summer, and he too had written the Kerr metric using transformed coordinates. But unlike the others, he anticipated the importance of this description of spacetime and, together with a colleague at another institution, published an important paper[1] that was to become the descriptor of what we now call the *Boyer-Lindquist coordinates*. Almost all of us who now use the Kerr solution begin by casting it into this form. Boyer, it seemed, was well on his way to becoming another major contributor to the revolution then in full swing.

But this is where the scientific story begins to jostle once again with the reality of a "shabby world." In the year 1963, when Roy Kerr and Maarten Schmidt were laying the foundation for the study of supermassive black holes, another individual, a troubled teenager by the name of Charles

---

1. See Boyer and Lindquist (1967).

Whitman, would introduce himself to the University of Texas as a mechanical engineering student.[2]

Over the next few years, Whitman's life would begin to career out of control, starting with a court-martial by the Marine Corps for gambling and possessing an illegal firearm. By 1966 he admitted depression to the university's doctor, who prescribed Valium and recommended psychiatric treatment. To the campus psychiatrist, Whitman expressed frustration with his parents' recent separation and his growing stress at work and in school. He was described by the campus doctor as "oozing hostility." In retrospect, this is hardly surprising, given that during the interview he revealed an urge to "start shooting people." Whitman never returned to the psychiatrist, later bemoaning the fact that he had found his first visit to "no avail."

Around 11:48 a.m. on August 1, 1966, Roy Kerr was sitting in George Ellis's office[3] when they heard the first cracking sounds coming from the mall area outside their building. Kerr believed that construction workers were busy with their activities, producing all sorts of noise as they went about their business. But Ellis's curiosity led him to open his office window, and he strained his neck to peer with squinting eyes at possible distant sources of the piercing sounds.

The previous day Whitman had purchased binoculars and a knife from a hardware store, and that evening began typing his suicide note: ". . . I don't quite understand what it is that compels me to type this letter. Perhaps it is to leave some vague reason for the actions I have recently performed. I don't really understand myself these days. I am supposed to be an average reasonable and intelligent young man. However, lately (I can't recall when it started) I have been a victim of many unusual and irrational thoughts. . . ."

The morning of August 1, shortly after 11:30 a.m., Whitman climbed to the observation deck of the main administrative building on the University of Texas campus, fatally wounding the receptionist along the way. The killing spree had begun. Within moments two families of tourists approached the deck attempting to look around, and Whitman began firing indiscriminately at the helpless individuals as they attempted to run back down the stairs. Some were killed instantly; others received permanent injuries.

The cracking sounds heard by Kerr and Ellis came minutes later, after Whitman had settled down on the outer deck and had begun firing at people on the mall (fig. 9.1). Like Kerr, many people on campus first dismissed

2. A detailed history of the events surrounding Charles Whitman during this period may be found in the Whitman Archives of the Austin History Center, at the Austin Public Library.

3. George Ellis—like Kerr, Penrose, Sachs, and Boyer—had been recruited to the Center for Relativity by Alfred Schild.

FIGURE 9.1. On the left is an image of the south mall on the University of Texas campus, as seen from the tower. In the center is a photo of the tower from which Charles Whitman shot scores of individuals. Physicist Robert Hamilton Boyer, on the right, was one of his earliest victims. (Images from the Charles Whitman Collection, AR.2000.002, Austin History Center, Austin Public Library.)

the sounds, not realizing that they were actually due to gunfire. But students started falling where they had previously stood, and seeing the carnage, at least one faculty member immediately called the Austin Police Department. Ellis quickly realized that being within eyesight of the tower was hardly safe, and he withdrew his head into the building, where together with Kerr, they remained for the rest of that afternoon's siege.

About ninety minutes after the shooting started, police officers learned of an employee at the university bookstore who was familiar with the interior of the tower, and deputized him to assist them in their attempt to reach the observation deck. Reaching the roof uneventfully, the police officers calmly stepped out and proceeded stealthily to the northeast corner of the deck. There they spotted Whitman curled up on the floor and fatally shot him, ending the siege.

His autopsy would later reveal that Whitman had a cancerous glioblastoma tumor in the hypothalamus region of the brain. Some would speculate that this abnormal cellular mass was pressed against the nearby amygdala, affecting his emotive passion. Neurologists wondered whether this medical condition was somehow responsible for his growing detachment from society and for the loss of control leading to his aggressive and violent behavior.

In the Center for Relativity, Kerr and his colleagues would not learn of the full extent of the tragedy until the following day, when it became apparent that someone they all knew very well had become one of Whitman's early casualties. Ironically, Robert Hamilton Boyer (fig. 9.1) had often complained of America's violent culture, and after serving as a postdoctoral fellow at the University of Texas, had decided to move his family back to England.

Boyer had just finished a one-month teaching assignment in Mexico and was on his way to Liverpool to begin a new appointment at the university there, where he could be nearer his pregnant wife, Lyndsay, and their children, Laura and Matthew. He was making a brief stopover in Austin to visit his friends and former colleagues and, just before noon that day, was heading out to lunch across the south mall with a companion. The fatal bullet struck him in the back, killing him instantly. Thus ended the life and promising career of a young relativist, who was present at the University of Texas during the exciting years when Kerr and his colleagues were revolutionizing our theoretical understanding of black holes.

Kerr and the others present that day would never fully recover from this traumatic event. Kerr wondered what the point was of engrossing oneself with Einstein's equations when such senseless chaos can so easily disharmonize an entire community. An air of despondency hovered over the Center for Relativity and the rest of the UT campus for the next several years.

10 :: NEW ZEALAND

The golden age of relativity lasted into the mid-1970s, though, in truth, most of the breakthroughs were realized by the end of the 1960s. Several of the "young Turks" in the relativity community, such as Stephen Hawking, would continue to define the role of Einstein's theory in physics and astrophysics for the remainder of the twentieth century. For Kerr, however, the exhilaration of 1963 would mark the zenith of his research career, imprinting with indelible ink his stamp on the history of science.

Today the Kerr metric is synonymous with black holes (chapter 11). A simple exploration on any Internet search engine uncovers millions of citations to this most famous solution of Einstein's equations. And yet a similar search for its discoverer's name, Roy Kerr, produces a relative paucity of references, numbering only in the tens of thousands. How did he become what one New Zealand journalist has called the "man of mystery"?[1]

Always a modest man, Kerr never sought the limelight. His approach to research was forged early in life, when he discovered the joy of solving problems for the sake of simply solving them. Throughout his career, his style has been to seek new challenges even before the one on which he might have been working had been met. Taking the time to write about his work did not come naturally. Even his landmark paper of 1963 was a simple one-and-a-half-page letter to the editor, when all of the elegant mathematics he had worked out to attain his solution should have filled forty to fifty pages of the main journal. Indeed, many other researchers who followed him would later write reports to "fill in" the information missing from Kerr's paper, when in fact much of that work had been completed and ignored by Kerr himself.

Kerr was fortunate in the 1960s to have found colleagues, such as Alfred Schild, who could fill this void—to urge him to put into writing all the work he had completed, sometimes in a frenzy. He relied on them heavily to chronicle the revolution that was taking place in relativity. In retrospect,

1. See the feature article by Marilyn Head in the New Zealand *Listener* (2004).

it is not difficult to understand that self-promotion was never part of his makeup.

Having reached such a pinnacle relatively early in his career, Kerr devoted the majority of his time seeking answers to other physical problems of similar impact, but this was becoming increasingly difficult in an era when most advances in relativity were rather abstract and not applicable to the real world. He found that most research in this area was becoming disappointingly fruitless.

It didn't help the situation that his solution drew instant attention from everyone working in relativity at the time. Recognizing its importance, many pursued it with enthusiasm (see chapter 8), motivated by the great discoveries they themselves might make in its trail. Barely a young man of twenty-eight when he published his landmark paper, and having come from a less intense lifestyle in New Zealand, Kerr was not prepared for the aggressiveness and adversarial manner of the scientific community in America. He had even felt obligated to wait for Ted Newman to complete his attempt at solving Einstein's equations (chapter 7) before he untethered himself psychologically to pursue it himself. Science generally doesn't work that way. When a problem is to be solved, many who have an interest pursue it unabashedly with vigor and motivation.

So it was quite a surprise to Kerr when he realized that others didn't share his view of how research ought to be conducted. And when one is reluctant to methodically publish one's own work, of course there will be instances when others produce papers of their own that broach the same subject.

From about 1966 to the end of the decade, Kerr gradually moved away from relativity and into more centralized topics in mathematics, including topology, and their possible relevance to physical systems. After all, what had started the golden age of relativity was the improvement in the mathematical technique used by the practitioners to find alternative approaches to solving complicated equations. But by then it was already beginning to look like most of what could be plumbed from relativity with pencil and paper that would be relevant to the real physical world had already been done. The golden age was in its waning days. Soon relativists would start to rely on detailed computer simulations to complete their work.

By the end of the 1960s, Kerr (fig. 10.1) was ready for a sabbatical—a break from the normal teaching duties of a university professor that permits him or her take an extended leave and visit colleagues elsewhere around the world. Ever since Kerr's arrival at the center in 1962, the relativity group had experienced a high rate of turnover in its personnel. Penrose himself left after only one year. Others, like Sachs, also arrived and then left again. A situation such as this can be good or bad, depending on the quality of the individuals coming through the institute. But without developing a sense of

FIGURE 10.1. In 1971 Roy Kerr returned to the University of Canterbury, where he would eventually lead the mathematics department for almost a decade. (Image courtesy of Roy Kerr.)

stability, a member of such an organization may feel a lack of certitude regarding the longevity of the enterprise. The Center for Relativity was itself having growing pains, and this seemed like a good time to change the scenery—at least temporarily.

However, Kerr's departure would in fact be permanent. He was to leave the United States just a year later, having been lured back to his homeland of New Zealand by the mathematics department at the University of Canterbury, where his odyssey began many years earlier. He revisited the Center for Relativity in Austin on several occasions afterward, but once his friend and colleague Alfred Schild died in 1977, Kerr's connection with the United States effectively came to an end.

Kerr set about giving back to his native New Zealand what it had given him in preparation for his voyage around the world, both literally and scientifically. He took on the role of head in the mathematics department in Christchurch and guided its growth to preeminence in the country.

But even though he was now preoccupied with departmental matters, and not fully aware that astrophysicists were finally embracing his work (see chapter 11), Kerr's reputation continued to grow overseas. Each succeeding black hole observation—within the galaxy and elsewhere—has provided new compounding evidence for the relevance of his solution to the real universe. As we shall see in the next chapter, today we find ourselves on the verge of retesting Einstein's theory, though this time in the strong-field

FIGURE 10.2. Roy Kerr (*standing on the right*) at the 1984 British Royal Society Awards ceremony, where he received the Hughes Medal. (Image courtesy of Roy Kerr.)

limit—that is, where the spacetime curvature becomes extreme. All previous tests of relativity have relied on measurements where the field is weak, leaving many to wonder whether this theory could still break down when gravity is superstrong. And the critical element in these tests will be the Kerr solution.

With the growing recognition of his work, Kerr has been the recipient of numerous awards, including the Hughes Medal from the British Royal Society (fig. 10.2). In the meantime, he has reawakened his desire to experience life's simple pleasures. Always a keen competitor, Kerr has become a master player of bridge. Winning twenty national titles after returning to New Zealand, he has represented his country many times at the world championship events, including the Bermuda Bowl in Venice. Then, in the early 1990s, he and his second wife, Margaret, sold their house and used the proceeds to buy a Cavalier 45 sailboat (fig. 10.3), one of the most seaworthy vessels in the world. Designed to handle rough seas, his intention was to sail around the globe, visiting friends and colleagues from relativity's golden age.

FIGURE 10.3. After his busy and influential years in the United States, Roy Kerr once again found leisure back home, which included boating in New Zealand's waters. He is shown here with his second wife, Margaret. (Image courtesy of Roy Kerr.)

But the unwritten rule of seafaring admonishes one to not set sail in a boat whose length is shorter than his age. Roy Kerr would now need a boat of enormous length, and thus he no longer sails. Instead, he has rediscovered his permanent roots in Christchurch, where his life began, as his lasting contributions to science begin to be recognized by the general public.

The Nobel Prize winner Ernest Rutherford has always been held in high regard. Ironically, Rutherford himself began his academic life at Canterbury College, the very school attended by Kerr in the mid-1950s. In fact, some of the classes attended by him were but an arm's length from what is now reverently known as Rutherford's Den—the basement laboratory where the famed physicist conducted his earliest experiments. But now New Zealanders have two physics heroes to revere. And in St. Andrew's Chapel (fig. 10.4), one can already see the beginnings of what will be Roy Kerr's enduring legacy in the country he brought to the world's stage during the halcyon days of relativity's golden age.

FIGURE 10.4. New Zealanders hold Roy Kerr and Ernest Rutherford in high regard for raising the profile of their country's scientific achievements on the global stage. In this image from 2007, Kerr sits before the stained-glass window in St. Andrew's Chapel, Christchurch, which features some of the great minds in physics, including Rutherford (*lower left*) and Kerr (*lower right*).

11 :: KERR IN THE COSMOS

Roy Kerr's return to New Zealand represented a significant turning point in his academic career. According to Thomas Wolfe, "you can't go home again," but resettling in Christchurch, a stone's throw from the school he attended as a youth, was for Roy a healthy change of pace.

Elsewhere, however, the pages of history continued to turn, and while new generations would infuse the sciences, the arts, and politics with their own deeds and aspirations, the seeds planted by Kerr and his colleagues during the golden age of relativity sprouted and grew with each new dramatic cosmic discovery.

Quite auspiciously, the cracking of Einstein's code and the discovery of objects for which this theory has much to say both occurred within the same year. And though Kerr's ten-minute presentation at the First Texas Symposium that fall went virtually unnoticed by all but the most ardent expositors of Einstein's theory, 1963 can rightly be acknowledged as the year in which supermassive black holes first began to capture the public's (and experts') imagination.

The Kerr solution is inextricably linked to these exotic objects because even though Einstein's theory of general relativity applies to everything in nature, its departure from Newtonian gravity—the overarching theory we are most familiar with in everyday life—clearly emerges in regions where gravity is so strong that even light cannot escape.

Since the 1960s, the uncloaking of supermassive black holes as the foundation for structure in the universe has been fostered by the rapid growth of space-based astronomy. Following the earliest rocket experiments (coincidentally also carried out in the 1960s), this development has been so explosive that it is no longer possible to catalog all the principal missions in a reasonably small space. And already, several Nobel Prizes have been awarded to key individuals involved in this breathtaking work, including Riccardo Giacconi in 2002 for his pioneering contributions to the discovery of cosmic X-ray sources (including black holes), and John Mather and George

Smoot in 2006 for their work with the cosmic background radiation created just after the big bang.

Many of the black hole discoveries have been made by the Chandra X-Ray Observatory, one of finest high-energy missions ever flown (fig. 11.1). Launched in July 1999, Chandra became the third in NASA's family of great observatories, which include the Hubble Space Telescope. Designed to sense photons with the same energy we encounter during a regular chest X-ray examination, Chandra's resolution is eight times better than that of any other X-ray telescope ever flown. With this capability, Chandra has provided us with highly detailed images of the sky, revealing objects radiating X-rays both near and far.

Indeed, one of the most spectacular images yet taken by Chandra (fig. 11.2) is that of the very same object that puzzled Maarten Schmidt back in 1963. We have learned that the quasar 3C 273's total X-ray power varies significantly in only a few hours. Astronomers can estimate the size of an object using the argument that nothing can travel from one end of the source to the other faster than light. Thus, if the light produced by 3C 273 changes in a matter of only hours, its size cannot be greater than a few light-hours across. This was quite a startling discovery because a source this small can easily fit within Neptune's orbit.

When we measure a quasar's total power output, we learn that it releases far more energy than an entire galaxy, yet the central engine that drives this powerful activity occupies a region smaller than our solar system. The otherwise preposterous notion that such a small volume is producing the same power as a hundred billion Suns makes it easy for us to identify objects such as 3C 273 as radiative manifestations of giant black holes.

Ironically, even the very first photograph ever made of a quasar, replicated here in this X-ray image (fig. 11.2), reveals far more than the black hole itself. Toward the lower right, a striking jet of plasma extends out from the core of 3C 273 into intergalactic space. Traveling at close to the speed of light, it flows along the black hole's spin axis, providing unmistakable evidence that the Kerr spacetime is at play near the central engine.

Today's supermassive black hole census includes some fifteen thousand distant quasars, though the total number is far greater. Due to their intrinsic brightness, the most distant among them are seen at a time when the universe was very young, a mere 1 billion years after the big bang.[1] The current distance record is held by an object that illuminates the cosmos some 13 billion light-years from Earth, meaning that it formed only 700 million years or so after the universe itself was created.

---

1. The best current estimate of the universe's age is approximately 13.7 billion years.

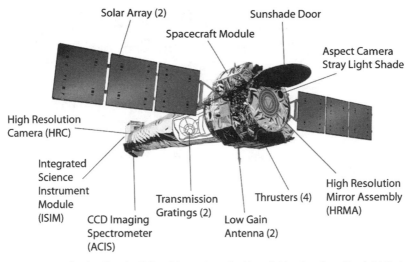

Solar Array (2)

Sunshade Door

Spacecraft Module

Aspect Camera
Stray Light Shade

High Resolution
Camera (HRC)

Integrated
Science
Instrument
Module
(ISIM)

CCD Imaging
Spectrometer
(ACIS)

Transmission
Gratings (2)

Thrusters (4)

Low Gain
Antenna (2)

High Resolution
Mirror Assembly
(HRMA)

FIGURE 11.1. On the Chandra X-Ray Observatory, the X-ray light enters from the right (just below the sunshade door) and is detected and measured at the rear of the facility where the CCD and other instrumentation are housed. The observatory is 45 feet long and 64 feet wide, from one edge of the solar array to the other. (Image courtesy of NASA.)

Quasars are good tracers of supermassive black holes, but they do not encompass all of these objects. Peering through several patches of sky relatively free of nearby stars (primarily those within our own galaxy), Chandra produced two of the deepest images ever made of the distant cosmos in X-ray light, one in the southern hemisphere, the other in the north. The latter is called the Chandra Deep Field-North and is shown in figure 11.3. Extrapolating from the number of suspected supermassive black holes in this image, we infer that at least 300 million of them must be dispersed across the universe.

However, the actual number of supermassive black holes must be even larger than this, since X-ray images such as figure 11.3 reveal only those objects whose orientation facilitates the transmission of their radiation toward Earth. Many more just happen to be radiating primarily in other directions, and we would have no direct knowledge of their existence.

Indeed, astronomers have known for many years that the medium between galaxies is pervaded by an X-ray haze. Unlike the cosmic microwave radiation left over from the big bang, the photons in this background glow are too energetic to have been produced during the universe's infancy. Instead, this radiation field suggests a more recent provenance associated with a population of black holes whose overall power dominates over everything else in the cosmos. Stars and ordinary galaxies simply do not radiate profusely enough to fit what we see toward the edge of the visible universe.

FIGURE 11.2. The quasar 3C 273 (the bright object in the upper left-hand corner) was one of the first objects to be recognized as a "quasi-stellar-radio-source" (quasar), due to its incredible optical and radio brightness. Insightful analysis led to the realization that 3C 273 is actually an incredibly powerful, distant object. This modern image, taken with the Chandra X-Ray Observatory, spans a region roughly 2,000 light-years across. (Photograph courtesy of H. L. Marshall, NASA, and MIT.)

It seems that for every suspected black hole, at least ten more are needed to produce this pervasive X-ray glow.

We should point out, however, that no one yet completely understands how these supermassive black holes produce their X-rays, or any photons for that matter. The black holes themselves do not produce light, other than possibly via the Hawking mechanism, but this process would account for a very faint glow, well below the threshold of detectability. Hawking radiation is predicted by the application of quantum mechanics to the behavior of a black hole's event horizon. The very notion of defining an event horizon makes sense as long as we can precisely place this surface and particles around it at perfectly known locations. But in quantum mechanics, there must always be some positional uncertainty, or an imprecision in energy and time, so general relativity, like all other classical theories, must acquire some minimal level of fuzziness.

FIGURE 11.3. The Chandra Deep Field-North X-Ray image. The vast majority of the five hundred or so sources in this view (corresponding to roughly 60 percent the size of the full Moon) are supermassive black holes. Extrapolating this number to the whole sky, one infers approximately 300 million such objects spread across the universe. (Image courtesy of D. M. Alexander, F. E. Bauer and W. N. Brandt, NASA and PSU.)

Quantum mechanics argues that we can never be entirely sure of a particle's position or its energy because in order for us to even know of its existence, we must disturb it to sense its presence. This uncertainty is the reason why physicists are uncomfortable with the idea of a perfectly localized and sealed event horizon, since these notions make no reference to the fuzziness of quantum mechanics on the smallest scales. In 1974 Stephen Hawking proposed the first step in the eventual resolution of the puzzle posed by the concept of an event horizon with quantum mechanical fuzziness.[2]

The name itself, *quantum* mechanics, reveals the essence of the physical description on a microscopic scale. Practically all of modern physics

2. Readers who would like to learn more about the technical aspects of this phenomenon, and the evaporation of black holes in general, will find the discussion in Thorne, Price, and Macdonald (1986) very helpful. See also Wald (1984).

is based on our ability to determine the value of certain quantities, such as the intensity of light, in terms of their location within a prescribed volume. Quantum mechanics tells us that at this level all such entities are to be thought of as comprising tiny bundles (or quanta) of "something," which in the case of light are known as photons. The connection between these bundles and the fuzziness we introduced earlier is that their size, energy, and lifetime are directly related to the scale of the imprecision, that is, how fuzzy the measurements of position or energy turn out to be.

Quanta such as photons may bubble up spontaneously out of the vacuum if an adequate source of energy lies nearby, but they always form in pairs, as if something must be split in order to create the fluctuation. The phenomenon discovered by Hawking[3] is directly associated with this creation of quantum particles in a vacuum due to fluctuations in the gravitational field of the black hole. Like a swarm of fireflies on a hot Georgia night, particles flash on and off just outside the event horizon. They live fleetingly and then annihilate with each other's counterpart to reestablish the vacuum after the fluctuation has subsided.

But here's the catch. The paired quanta produced in this fashion annihilate very quickly (in about one-millionth of a millionth of a millionth of one second). Some pairs, argued Hawking, will have a member that dips below the horizon, abandoning its partner to the whim of the outside universe. Without a partner to annihilate, the detached particle flees the black hole's sphere of influence and merges into the flux of escaping radiation headed for infinity. To us, this looks like the black hole is actually radiating, though the mechanism is clearly indirect. Hawking had explored this process for the collapse of stars into black holes, reasoning that quantum particle creation would be significant during the rapid collapse due to the changing gravitational field. Expecting the emission of these particles to subside once the star had reached the singularity, he was surprised to find that instead the emission reaches a steady rate dependent on the photon's energy.

The wavelength of the radiation produced by the Hawking mechanism is commensurate with the size of the black hole that creates it. Such large wavelengths translate into a very feeble photon energy indeed, so although black holes may not be completely black after all, they would nonetheless be very, very faint objects, if this were the only means of radiating available to them.

But black holes do not exist in isolation. At least the ones we know about

---

3. Some of Hawking's early discussion on this topic appeared in a paper published by *Nature* in 1974.

lurk in the nuclei of active galaxies and are therefore surrounded by streaks of gaseous plasma. Occasionally, some of this matter winds its way inward and becomes haplessly trapped in the deep gravitational well, which compresses and heats it during its inexorable journey toward the event horizon, sometimes reaching temperatures as high as a billion degrees before disappearing from sight. It is apparently this hot plasma that produces the X-rays we see in images such as figures 11.2 and 11.3.

When astronomers add up all the radiative power produced by this enormous population of supermassive black holes, they infer that about *half* of all the universe's radiation produced after the big bang is due to them. Ordinary stars no longer monopolize the power as they had for decades prior to the advent of space-based astronomy, which has finally allowed us to see the true nature of structure in the universe.

Kerr's solution to Einstein's equations of general relativity describes each and every one of these gargantuan pits of darkness because they are all spinning. Some spin faster than others, so their internal structure varies from one object to another (see fig. 8.6). Probing deeply into a quasar's core to see this structure is still beyond our reach, but the situation is certainly changing rapidly insofar as the center of our own galaxy is concerned (see fig. 12.1).

We know supermassive black holes are spinning because of the appearance of objects such as Cygnus A (fig. 11.4) and 3C 273 (figure 11.2). Located 600 million light-years from Earth, the luminous extensions in the former cast an aspect only one-tenth the diameter of the full Moon. But we know that in reality they project out from the black hole an incredible distance three times the size of the Milky Way.

The process responsible for producing the pencil-thin jets in Cygnus A must be stable for at least as long as it takes the streaming particles to journey out from the center of the galaxy to the extremities of the giant lobes. Given the overall size of this structure, it would have taken these particles (traveling at the speed of light) almost 1 million years to get there.

A spinning black hole is required to accomplish this feat since the axis of its spin functions as a gyroscope, whose direction determines the orientation of the jets. Astronomers are still not certain exactly how the expulsion of matter occurs, but almost certainly the twisting motion of the swirling Kerr spacetime near the black hole's event horizon is the cause. The Kerr solution, which describes the dragging of inertial frames about the black hole's spin axis, provides a natural setting for establishing the preferred direction of this ejective process.

But when all is said and done, do black holes really contain a ring singularity at their core? Kerr's solution to Einstein's equations corresponds to a

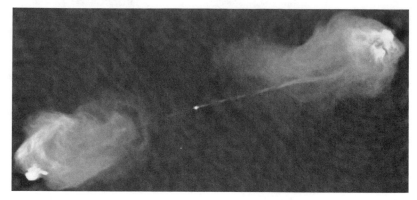

FIGURE 11.4. A radio telescope image of the powerful central engine and its relativistic ejection of plasma in the nucleus of Cygnus A. The highly ordered structure spans a region over 500,000 light-years in extent, fed by ultra-thin jets of energetic particles beamed from the compact radio core between them. The giant lobes are themselves formed when these jets plow into the tenuous gas that exists between galaxies. Despite its great distance from us (over 600 million light-years), Cygnus A is still by far the closest powerful radio galaxy and one of the brightest radio sources in the sky. The fact that the jets must have been sustained in their tight configuration for over half a million (possibly as long as 10 million) years means that a highly stable central object—probably a rapidly spinning supermassive black hole acting like an immovable gyroscope—must be the cause of all this activity. (Photograph courtesy of Chris Carilli and Rick Perley, NRAO, AUI, and NSF.)

highly idealized set of conditions, including the assumption of an asymptotic limit in which spacetime actually ceases to change with time. This was a necessary starting point in order to produce a simplified set of equations yielding a manageable solution (see chapter 7).

The awkwardness of this result arises from the fact that the internal volume of a Kerr black hole, depicted in figure 8.6, must be a vacuum. Even Roy Kerr himself is skeptical that such a configuration can ever be reached. Almost certainly, the interior contains some form of mass energy (or even simply energy, if mass as we know it is not permitted there), which would ruin the beautiful structure of the Kerr black hole's simplified interior. As noted earlier (chapter 8), closed timelike loops associated with the ring may in principle permit time travel, but, ironically, the moment an observer is brought into the mix, mass and/or energy is introduced there as well, and the simple structure disappears.

We may never know what the actual interior of a black hole looks like, but given this uncertainty, astronomers are left wondering how seriously the *real* (exterior) spacetime surrounding a spinning black hole departs from the idealized one we have been considering here. These days, experimental black hole research is attempting to develop the means to probe the environment just outside an event horizon. Arguably the most spectacular of these will be the actual imaging of a black hole, a prospect that appears to be within reach in at least one case (see chapter 12).

But there are other, less direct, methods as well—one based on "hearing" a black hole. Matter falling into one of these objects is always moving, its path often coinciding with an orbit like that of a planet around the Sun. Stressed by the intense gravity near the black hole's event horizon, this material is cajoled into radiating away the energy it gains on its inward journey. And as the gas winds its way through the swirling spacetime back and forth around the dark pit, it produces an outward flow of photons with a characteristic beat corresponding to the period of its revolution.

For example, in our solar system, Mercury would produce a high-pitched "sound," given that its period around the Sun is relatively short compared to that of the other planets. Earth would be somewhere in between, and Pluto—the farthest and the slowest—would produce the deepest bass. By identifying a planet's orbital pitch, one could easily determine its distance from the Sun. In a similar fashion, by measuring the "sound" produced by gas orbiting about the black hole, astronomers can infer its distance from the singularity.

And it is this mapping ability that X-ray astronomers are now exploiting to determine the orbital trajectory of gas falling into supermassive black holes, but also much smaller objects, five to twenty times the mass of the Sun, formed in supernova explosions.

Gas captured by the black hole is eventually absorbed through its event horizon. However, the allowed proximity of its orbit is entirely based on the spacetime fabric. The frame dragging around a spinning object can provide additional support to the radiating plasma, allowing it to orbit closer to the point of no return than it could otherwise manage. By "listening" to the beat of this matter, astronomers can measure exactly how close it gets before finally falling in.

In principle, one can therefore produce a map of the spacetime surrounding a black hole for direct comparison with the predictions of the Kerr solution. Indeed, with sufficiently precise data, one can even use this method to accurately measure the black hole's spin.

Already, astronomers at the Goddard Space Flight Center in Maryland have demonstrated the feasibility of this work, by showing that the X-ray pulsations from a microquasar 10,000 light-years from Earth reveal an emitting region orbiting closer than 49 kilometers from the center of the black hole.[4] As its name suggests, a microquasar has many of the attributes of a "real" quasar such as 3C 273 (fig. 11.2), though much smaller in scale—

4. See Strohmayer (2001).

and correspondingly much closer to Earth. Whereas quasars are billions of light-years away, microquasars (roughly ten times the mass of the Sun) reside within our own galaxy, at a distance of only 20,000 light-years or so.

The Goddard scientists used the Rossi X-Ray Timing Explorer, launched by NASA in December 1995, to record the rapid fluctuations coming from this object. They found unique patterns in the X-radiation caused by hot plasma dancing back and forth on an orbit so close to the black hole that the material could not have survived there without the support provided by the whirlpool-like rotation of a Kerr spacetime.

We find ourselves barely at the dawn of black hole research. Though we now have confidence that Einstein's code has been cracked, it is not yet clear how much more can be done, but surely there is more, even without worrying about the marriage of general relativity with quantum mechanics. If the Kerr solution is not the final answer, then "listening" to black holes and photographing them will certainly help us reach the next level.

AFTERWORD

Einstein's theory of general relativity had several remarkable successes in the early part of the twentieth century. It explained certain anomalies in the way Mercury orbits the Sun, and its prediction that light would be gravitationally attracted by the Sun's gravitational field was spectacularly verified in 1920. In that same decade, it was used to explain why the light from distant galaxies appeared redshifted: the universe is expanding and therefore galaxies are moving away from us. However, relativists made little progress after this initial period, and Einstein's theory was largely ignored by mainstream physics for many years.

After the Second World War, this all started to change. First of all, many very good physicists and mathematicians became interested in general relativity, partly because of the need to relate it to quantum theory and partly because of the influence of several outstanding scholars who encouraged younger scientists to work in that area. As astronomers learned more about the cosmos, particularly its evident expansion and the growing number of unusual objects that populated it, theoreticians began to suspect that a post-Newtonian theory of gravity was essential to an eventual explanation of the universe.

By the summer of 1963, Maarten Schmidt at Caltech had shown that certain starlike objects (now called quasars) were actually distant objects emitting enormous amounts of energy. Nobody understood how they could be so bright. In an effort to unravel this mystery, several hundred astronomers, astrophysicists, and general relativists gathered for a conference in Dallas, held in early December of that year. This would be the First (of what since then has become the biennial) Texas Symposium on Relativistic Astrophysics.

By a rather strange coincidence, I had just found the correct theoretical solution to Einstein's equations describing what we now call spinning black holes. But when I presented this at the conference, the astrophysicists did not even bother to listen, much to the annoyance of the relativists present.

How times have changed! It was realized within only a few years that massive black holes were the engines driving quasars, and many now suspect that these objects may have been crucial to the formation of structure in the universe.

I myself followed a rather unorthodox path getting interested in the spacetime surrounding spinning black holes. By all accounts, I was a precocious youngster, placed in classes with older children at school but essentially left to my own devices for mathematics. Resources were so scarce in the immediate postwar years that my secondary school could not find anyone to teach either mathematics or physics. I remember going through the mathematics book on my own, doing all the exercises, and then being told to simply go through it again for want of anything better to do.

This did not matter very much since I entered the University of Canterbury in Christchurch at the age of sixteen and was allowed to skip the first two years of mathematics and go straight into the third-year courses. I could have passed the master's exams the following year, but because of university rules, I had to wait for almost three years before I could move to Cambridge for graduate studies. This is not to question the good intentions of people there, but simply to say that the New Zealand of the 1950s was just not set up for promoting the intellectual development of its youth in the manner practiced by the established universities in the northern hemisphere. I should have studied modern mathematics and physics, but there was no money for new books and therefore the library was a carryover from the nineteenth century. Nobody was doing any theoretical research in either mathematics or physics at that time in Christchurch, so I spent my last two years there mostly playing golf, tennis, badminton, and finally bridge.

When I moved into the mainstream of academic life in England, my old habits and style were hard to break. I had by then developed somewhat of a maverick streak, a lack of attachment to the traditional ways of doing things, and—truth be told—a fairly mischievous skepticism of what I read in books and heard from elders. The first half of my time at Cambridge was spent studying modern mathematics and physics. My supervisor then left Cambridge, and I spent my last year and a half without a supervisor. It was during this time that I wrote a thesis on the motion of orbiting bodies within the context of Einstein's theory.

Most people credit their mentors for their success in life. Fortunately, I have had no mentors and so have always been able to think for myself. A mentor at Cambridge would have convinced me that I was wasting my time even looking at this problem, since everything had apparently already been done by those who had worked on it during the previous thirty-five years.

For sure, there is a downside to not having a mentor to look after you as you grow in your profession. There is some truth to the old adage that "it is not just what you know, it is also who you know that matters." Science has its cliques just like all other areas of life. The best advice I can give to a young student is to choose your adviser and cliques wisely.

However, there is an upside too—one that in the long run made all the difference for me. Untethered to any established ideas, I felt free to question what I was told, to criticize anything and everything that I believed was wrong, and to pursue paths that others might have shunned. It was precisely this intellectual flexibility that allowed me to discover the mathematical expression for the spacetime surrounding a spinning object.

In 2005 a conference was held in Christchurch to celebrate my seventieth birthday. Fulvio Melia came and gave a talk in which he outlined all the strange and wonderful phenomena that astrophysicists now associate with rotating black holes. This was undoubtedly the best talk to a general audience that I have ever heard. We became very good friends, and I was delighted when Fulvio came to the University of Canterbury in 2007 to write this book. It is a remarkable piece of writing, capturing beautifully the period we now refer to as the golden age of relativity. I hope you enjoyed reading this story as much as I enjoyed "participating" in it over the years.

ROY KERR
Christchurch, New Zealand
June 2008

REFERENCES

Armstrong, K. 2004. *Buddha*. New York: Penguin.

Boyer, R. H., and R. W. Lindquist. 1967. "Maximal Analytic Extension of the Kerr Metric." *Journal of Mathematical Physics* 8:265–81.

Bromley, B., F. Melia, and S. Liu. 2001. "Polarimetric Imaging of the Massive Black Hole at the Galactic Center." *Astrophysical Journal Letters* 555:L83–L87.

Byers, N. 1999. "E. Noether's Discovery of the Deep Connection between Symmetries and Conservation Laws." *Israel Mathematical Conference Proceedings* 12:40–51.

Carter, B., and M. S. Carter. 2006. *Simon Newcomb: America's Unofficial Astronomer Royal*. St. Augustine, FL: Matanzas Press.

Chandrasekhar, S. 1987. *Truth and Beauty: Aesthetics and Motivations in Science*. Chicago: University of Chicago Press.

Corry, L., J. Renn, and J. Stachel. 1997. "Belated Decision in the Hilbert-Einstein Priority Dispute." *Science* 278:1270–73.

Cox, D., and E. J. Flaherty. 1976. "A Conventional Proof of Kerr's Theorem." *Communications in Mathematical Physics* 47:75–79.

Damour, T. 1987. "The Problem of Motion in Newtonian and Einsteinian Gravity." In *Three Hundred Years of Gravitation*, pp. 128–98, edited by S. W. Hawking and W. Israel. Cambridge: Cambridge University Press.

Debney, G. C., R. P. Kerr, and A. Schild. 1969. "Solutions of the Einstein and Einstein-Maxwell Equations." *Journal of Mathematical Physics* 10:1842–54.

Drake, S. 1981. *Cause, Experiment, and Science*. Chicago: University of Chicago Press, 1981.

Dyson, F. W., A. S. Eddington, and C. Davidson. 1920. "A Determination of the Deflection of Light by the Sun's Gravitational Field, from Observations Made at the Total Eclipse of May 29, 1919." *Philosophical Transactions of the Royal Society of London*, 291–333.

Einstein, A. 2006. *Relativity: The Special and General Theory*. New York: Penguin Classics.

Einstein, A., L. Infeld, and B. Hoffmann. 1938. "The Gravitational Equations and the Problem of Motion." *Annals of Mathematics* 39:65–100.

Eliade, M. 1984. *A History of Religious Ideas*, Vol. 2, *From Gautama Buddha to the Triumph of Christianity*. Chicago: University of Chicago Press.

Falcke, H., F. Melia, and E. Agol. 2000. "Viewing the Shadow of the Black Hole at the Galactic Center." *Astrophysical Journal Letters* 528:L13–L17.

Ginzburg, V. L. 1964. "Experimental Verification of General Relativity Theory." In *Conférence Internationale sur les Théories Relativistes de la Gravitation*, edited by L. Infeld, pp. 55–69. Oxford: Pergamon Press.

Gold, T. 1965. "Summary of After-Dinner Speech." In *Quasi-Stellar Sources and Gravitational Collapse*, edited by I. Robinson, A. Schild, and E. L. Schucking, p. 470. Chicago: University of Chicago Press.

Guthrie, W. K. C. 1981. *A History of Greek Philosophy*. Cambridge: Cambridge University Press.

Hawking, S. W. 1974. "Black Hole Explosions?" *Nature* 248:30.

Hawking, S. W., and G. F. R. Ellis. 1973. *The Large Scale Structure of Spacetime*. Cambridge: Cambridge University Press.

Head, M. 2004. "Man of Mystery." *New Zealand Listener* 195:1–3.

Hentschel, K. 2005. "Testing Relativity." In *Physics Before and After Einstein*, edited by M. Mamone Capria, pp. 163–80. Amsterdam: IOS Press.

Infeld, L. 1964. "Opening of the Conference." In *Conférence Internationale sur les Théories Relativistes de la Gravitation*, edited by L. Infeld, pp. xv–xvi. Oxford: Pergamon Press.

Israel, W. 1967. "Event Horizons in Static Vacuum Space-Times." *Physical Review* 164:1776–79.

Israel, W., and E. Poisson. 1990. "Internal Structure of Black Holes." *Physical Review D* 41:1796–1809.

Kerr, R. P. 1963. "Gravitational Field of a Spinning Mass as an Example of Algebraically Special Metrics." *Physical Review Letters* 11:237–38.

———. 1965. "Gravitational Collapse and Rotation." In *Quasi-Stellar Sources and Gravitational Collapse*, edited by I. Robinson, A. Schild, and E. L. Schucking, pp. 99–102. Chicago: University of Chicago Press.

Kerr, R. P., and A. Schild. 1965. "Some Algebraically Degenerate Solutions of Einstein's Gravitational Field Equations." *Proceedings of Symposia in Applied Mathematics* 17:199–209.

Kohn, L. 2005. *Daoism Handbook*. Boston: Brill Academic Publishers.

Lederman, L. M., and C. T. Hill. 2004. *Symmetry and the Beautiful Universe*. Amherst, NY: Prometheus Books.

Lense, J., and H. Thirring. 1918. "On the Influence of the Proper Rotation of Central Bodies on the Motions of Planets and Moons According to Einstein's Theory of Gravitation." *Physikalische Zeitschrift* 19:156–63.

Magueijo, J. 2003. *Faster than the Speed of Light: The Story of a Scientific Speculation*. Cambridge, MA: Perseus, 2003.

Melia, F. 2001. *Electrodynamics*. Chicago: University of Chicago Press.

———. 2003. *The Edge of Infinity—Supermassive Black Holes in the Universe*. Cambridge: Cambridge University Press.

———. 2006. *The Galactic Supermassive Black Hole*. Princeton, NJ: Princeton University Press, 2006.

Moffat, J. 1993. "Superluminary Universe: A Possible Solution to Initial Value Problem in Cosmology." *International Journal of Modern Physics D* 2:351–66.

Pound, R. V., and G. A. Rebka Jr. 1960. "Apparent Weight of Photons." *Physical Review Letters* 4:337–41.

Price, R. H. 1972. "Nonspherical Perturbation of Relativistic Gravitational Collapse." *Physical Review D* 5:2419–38.

Robinson, I., and A. Trautman. 1964. "Exact Degenerate Solutions of Einstein's Equations." In *Conférence Internationale sur les Théories Relativistes de la Gravitation*, edited by L. Infeld, pp. 107–14. Oxford: Pergamon Press.

Russell, B. 1996. *The Principles of Mathematics*. New York: W. W. Norton.

Sachs, R. K. 1964. "The Characteristic Initial Value Problem of Gravitational Theory." In *Conférence Internationale sur les Théories Relativistes de la Gravitation*, edited by L. Infeld, pp. 93–105. Oxford: Pergamon Press.

Schmidt, M. 1963. "3C 273: A Star-Like Object with Large Red-Shift." *Nature* 197:1040.

Shapiro, S. L., and S. A. Teukolsky. 1991. "Formation of a Naked Singularity—The Violation of Cosmic Censorship." *Physical Review Letters* 66:994–97.

Strohmayer, T. 2001. "Discovery of a 450 HZ Quasi-Periodic Oscillation from the Microquasar GRO J1655-40 with the Rossi X-Ray Timing Explorer." *Astrophysical Journal Letters* 552:L49–L53.

Thorne, K. S., R. H. Price, and D. A. Macdonald. 1986. *Black Holes: The Membrane Paradigm*. New Haven, CT: Yale University Press.

Wald, R. M. 1984. *General Relativity*. Chicago: University of Chicago Press.

Weinberg, S. 1972. *Gravitation and Cosmology: Principles and Applications of the General Theory of Relativity*. New York: Wiley.

INDEX

Thorne, Kip, 37, 115, 131
time dilation, 13, 28–32, 38
time travel, 95, 99, 118
topology, 89, 106
Torrence, R., 86
Townes, Charles, 32
Trautman, Andrzej, viii, 44–47, 52, 70–71, 130
Trinity College, 60
trou noir, 93
two horizons, 93–94

unified field theory, 63–64, 67
universal expansion, 89, 125
University of California, Santa Barbara, 54, 68
University of Texas at Austin, 1, 4, 54–55, 68, 76, 100–103
    tower massacre, 100–104
USS *United States*, 55

vacuum solution, 84, 87, 96, 130
variable speed of light, 64
VIRGO, 37, 42, 79

Warsaw, 39, 48–51, 67, 123
wavelength, 30–32, 38, 80, 116
Weiss, Rainer, 37
Wheeler, John, 66, 74, 82, 86–95
Whitman, Charles, 102–4
Whitman Archives, 102
Wiltshire, David, viii
Wolfe, Thomas, 111
Wright-Patterson Air Force Base, 54, 66, 76

X-ray background, 113
X-ray telescope, 112

Zeno, 4–14, 123
Zeus, 7